YOU'RE A MIRACLE
MIRACLE

(AND A PAIN IN THE ASS)

YOU'RE A MIRACLE

MIRACLE

(AND A PAIN IN THE ASS)

Embracing the Emotions,
Habits, and Mystery
That Make You You

MIKE McHARGUE

CONVERGENT

NEW YORK

Published in the United States by Convergent Books,
an imprint of Random House, a division of
Penguin Random House LLC, New York.

CONVERGENT BOOKS is a registered trademark and
its C colophon is a trademark of Penguin Random House LLC.

LIBRARY OF CONGRESS CATALOGING-IN-PUBLICATION DATA
Names: McHargue, Mike, author.
Title: You're a miracle (and a pain in the ass) / Mike McHargue.
Description: New York: Convergent, [2020] | Includes index.
Identifiers: LCCN 2019048567 (print) | LCCN 2019048568 (ebook) |
ISBN 9781984823243 (hardcover) | ISBN 9781984823250 (ebook)
Subjects: LCSH: Self-acceptance. | Self-realization. |
Self-management (Psychology)
Classification: LCC BF575.S37 M43 2020 (print) |
LCC BF575.S37 (ebook) |
DDC 158.1—dc23
LC record available at https://lccn.loc.gov/2019048567
LC ebook record available at https://lccn.loc.gov/2019048568

Printed in Canada

crownpublishing.com

2 4 6 8 9 7 5 3 1

First Edition

Book design by Victoria Wong

For Rachel Held Evans, who helped me
(and millions of others) find home

In times of stress, the best thing we can do for each other is to listen with our ears and our hearts and to be assured that our questions are just as important as our answers.

—FRED ROGERS, *The World According to Mister Rogers*

Contents

8. The Forest and the Trees
   Why Learning to Live in Our Bodies
   Might Save the World

Author's Note

(You Don't Want to Skip This One)

Dear Reader,

Last spring, I watched a butterfly emerge from its cocoon in my front garden. I was struck by how it struggled. The tiny creature often stopped, presumably out of exhaustion, even on this critical threshold of its metamorphosis from caterpillar to a creature that can soar through the air. At times, I was concerned it wouldn't make it at all.

Forgive me if this seems presumptuous, but I think I know how the little bug felt. In many ways, this book is my own cocoon, spun from words and paper instead of silk. A couple years ago, when I started writing it and exploring the science of why we struggle to change our behavior, I was self-assured in a peculiar way: finding confidence not in my mastery of life, but in my capacity to surrender to it. But life has a way of challenging the stories we tell ourselves to survive, and the past eighteen months have been a metamorphosis for me. I have been through loss and grief, a medical diagnosis, and a mental health crisis, and lost all the money I spent a lifetime saving. If you're going to read this book, you and I will spend many hours together—and for that to work, I have to be honest with you at the beginning.

This is a dangerous book. The process of researching and applying its insights—from psychology and sociology to neuroscience and behavioral economics—changed my life in dramatic and unexpected ways. There has been pain, struggle, and loss. In the pages that follow, I'm going to be honest about what I've learned, in the hope that it might help anyone else who's fighting to emerge from their own cocoon.

Parts of this book could be a tough read, depending on how your life has unfolded. I write about my experiences with trauma and suicidal thoughts, and if you've got some scar tissue, my stories may bring up pain. (You'll find that the opening of the book sets the stage for honesty.) Take a break if you need it. You can set this book down anytime you'd like. I've included ample resting points in the flow of the text to help. I'll be here, in these pages, whenever you need me, so there is no need to rush.

Finally, while the stories I tell in this book are based on real events, I've sometimes reordered them, combined separate events together, and offered my recollections of what other people said, all for the sake of time and readability. Consider them "based on a true story," as they say across town in Hollywood.

I'm so excited to share this book with you.

Peace, Love, Entropy,
Mike McHargue, or "Science Mike"
Los Angeles, California, 2019

YOU'RE A MIRACLE

(AND A PAIN IN THE ASS)

Why Did I Do That?

The Battle of You vs. You

The barrel of a shotgun tastes like pocket change and fireworks. The flavor is overwhelming, like a battery pressed to your tongue, though the taste and aroma are quickly matched by the discomfort of what it takes, physically, to put such an instrument in your mouth.

After placing the stock of the weapon on the ground, you lean forward and awkwardly bow your head, as if in prayer. If your goal is suicide—and it must be, or why else are you pointing a shotgun at your head?—you want to make sure as much shot as possible passes through your brain. So, still leaning forward, you lift your forehead into a less penitent posture. The result is pain, with the barrel smashing against your teeth and jabbing into the gums behind them.

Shotguns are called "long guns" for a reason. With the barrel in your mouth, you can't reach the trigger while maintaining the critical head-to-barrel angle. My solution to this problem was improvised: I slipped off my right shoe and put my toe on the trigger. Doing so stretched my hip flexors uncomfortably and put strain on my knee. I've never liked pain, but I could cope with this discomfort. In a few seconds, I wouldn't feel anything ever again.

I was sixteen years old, and tired of pain, rejection, and fear. My heart ached so much, and so constantly, that I didn't want to have a heart anymore. So here I was, sitting in my parents' bedroom with my father's hunting gun in my mouth. I took a deep, smoky, metallic breath and pushed the trigger with my toe. . . .

I FELT THE *clunk* even more than I heard it. It reverberated through my teeth and into my skull.

I'll never forget that feeling, the ultimate anticlimax. Confusion washed over me, bordering on panic. How was I alive? Turns out, I'd cocked the gun successfully but hadn't loaded it right. I was saved by my ignorance of guns, and too afraid to try again. My heartbeat hit triple digits as I fell on the brown carpeted floor and sobbed.

I hadn't been an obvious candidate for a suicide attempt. I was a lanky teenager who played in a band and had a lot of friends. My parents loved me and provided a lifestyle that was comfortably middle class. I had a car and everything. My life looked about as stable and secure as adolescence can be.

Lying on the floor that day, I couldn't believe that I'd pulled the trigger. Taking one's own life goes against some of our most powerful instincts, those for living and self-preservation. This is why, when someone attempts suicide, the most obvious response for the people closest to them is to ask the unanswerable question, "Why?" For people like me, however—those who have survived suicide attempts—there's an additional, equally puzzling question: Did I really want to die?

The answer seems clear. After all, I'd put a loaded gun in my mouth and pulled the trigger. But if I'd really wanted to die, why hadn't I picked up the shotgun, troubleshooted the problem, and then tried again? How could I want to die badly enough to pull a trigger once, but not twice?

Why, instead, did I lie on the floor, full of grief and shame, lamenting that I had failed even at killing myself, while feeling thrilled by the tears on my cheeks because those tears meant I was still alive? How could I be happy at the same time I wished to be dead?

My struggle over suicide was life or death. My feelings warred with themselves, fighting in equal measure for another breath and a final one.

But who was I struggling with? I was the only one in the room.

ON THE MORNING I write this, the California sun streams into the bedroom I share with my wife, Jenny. Our dogs have started to prance around the foot of our bed, anxious for us to wake up. They're older dogs, with gray muzzles and a love of naps. They're mellow enough to not bark or jump on the bed, but they've learned that their nails make a ceaseless *click-clack* on the hardwood floor that reliably breaks the slumber of their human roommates. "Time to get up," they say, as politely as a canine can muster.

This year has brought a lot of change to my family. One of the smaller changes was moving across the country from Florida to California. Jenny and I had both lived in Tallahassee our entire lives, and it's hard to imagine a bigger change than moving from our sleepy southern hometown to a city like Los Angeles. The move has been both thrilling and exhausting. A mega-city has a lot more things to do, but somehow far fewer places to park.

I often cope with stress by eating, and L.A. is more than happy to enable me. I've been packing on the pounds while at home—and packing them on while I travel for events. It's a bad cycle. So last night, I decided that I would take a walk when I woke up.

I love going for walks. Aside from the obvious health benefits, I find that time spent walking outside pays dividends in my emotional well-being, and makes me more creative and focused once I sit down at my desk. Plus, the perfect Southern California weather knocks out my favorite excuse for avoiding exercise: heat combined with high humidity. Most mornings, I wake with the intent of roaming around the foothills of the San Gabriel Mountains before getting on with my day. But this morning, I just watched that famous golden sunlight through the window and ate a banana in my favorite chair, even as a small voice in my head told me, "It's not too late! Just go outside and walk now."

This happens to me all the time. How many mornings

have I wasted by snagging my iPhone off of the night-stand, instead of leaving it there until I've eaten breakfast with my family? I can't even estimate the percentage. Why do I always walk past the juice bar and buy my afternoon snack at the little shop Donut Friend?

Why do I so often choose to do things I know aren't best for me? Why do I sit on the couch when I love to go outside? Why do I watch late-night TV when mornings are my favorite part of the day? I don't even like television. Why do I do these things even though I know I'll feel shame later, after I do them?

Lucky for me, I am not alone. Part of my work (we'll get to this later) involves listening to people talk about the parts of their lives that frustrate them most. It seems we all struggle with late-night Netflix binges, skipped workouts, and deeper existential concerns about our mental health and belonging. I've learned that most people are fighting their own version of this internal war. What's good for us later fights what feels good right now. Guiding this internal debate is a set of hazy, indistinct emotions and impulses that drives us to do things we have little interest in, or feel guilty about later.

My own feelings and desires often confuse me, or seem in outright rebellion against my will. But it's not just that I so often struggle to do what I know is right. Sometimes I can't understand why I did something at all.

Turns out, I am not the first person to study this struggle.

• • •

IMAGINE YOU'VE BEEN invited to participate in an experiment that will test your language skills. The lab is at the end of a hallway, so you walk down it and open the door. A research assistant greets you inside and explains that the test is quite simple: You'll be given groupings of five words and asked to create sentences using four of them. Once you're done, you're told to notify her that you've finished.

It's an easy task, familiar to anyone who has ever made poetry with magnets on the fridge. The cards labeled "woman" and "kind" connect with "the" and "is" to form "the woman is kind." The test is so easy, you may wonder if there are more instructions to come. But when you show your work to the research assistant, she thanks you and tells you that you've successfully completed the test.

So you leave, walking down the same hallway you walked in from, and get on with your day. No big deal, right?

Well, what if I told you that the speed at which you walked down that hallway was predicted by the words that the researchers gave you in your word scramble? What if I told you that when words we associate with elderly people—like "wise," "obedient," "courteous," or "alone"—are included, subjects in the experiment walk as much as 10 or 20 percent slower leaving than they did when they came in?

It seems unbelievable, doesn't it? But that's exactly what the BBC found when they tried to reproduce a famous experiment on a well-known concept in psychology called "priming." There have been many variations of this study, and they've produced some truly shocking results.

Let's change the experiment. When you're done with your word scramble, the research assistant engages you in a deep, important-sounding conversation that drags on a little too long. Would you interrupt her to tell her you've completed your task? Researchers found that when people's word scrambles incorporated words we associate with being polite—like "kind" or "respected"—they usually waited a full ten minutes for the conversation to finish. But when people got a word scramble loaded with rude words, like "brazen" or "aggressively," they waited only a minute or two before interrupting and saying, "Hey, I'm done here."

Experiments in priming have made people perform better at Trivial Pursuit after completing word scrambles involving words associated with intelligence. Another study indicated that people are more likely to vote for a tax increase to support education if their polling place is a school.

The point is this: We feel like we're in control of our lives, and that we make decisions based on evaluating circumstances with our rational minds. But research is showing us that we're far from the rational actors we believe ourselves to be. A host of factors are at play in every thought or feeling we have, and every action we undertake.

I imagine some of your minds are blown by the concept of priming, while others of you may have heard about it in a TED talk or read about it in a bestselling book. But I choose to mention it here because it sets an important foundation for the rest of this book: You are not the rational, autonomous actor that you believe yourself to be. Not even close.

We experience our lives as a series of choices that we make, but when we examine human behavior via behavioral science or brain imaging, we find that our choices don't happen in a vacuum. Covert cues and features of our biological systems act like the strings on a marionette, guiding our movements with inexorable tugs toward being courteous or rude—or even conservative or liberal. The sciences have long revealed the limits of our willpower and decision making—yet when it comes to addressing the problems that frustrate us most, our politics, our religions, and an endless library of self-help tomes all place an overwhelming emphasis on individual responsibility and initiative. Something has to give.

I MAKE MY living hosting two podcasts: *The Liturgists Podcast* and *Ask Science Mike*. In the course of making those programs, I get asked thousands of questions every year from people struggling to understand their own beliefs, emotions, and behaviors. Those questions are the impetus behind this book. They read like mini-biopics, all of them pointing to our most common human dilemma.

In Boise, a college student asks how she can accept that climate change is real, while also eating beef and buying things from Amazon Prime all the time. She understands that *her own activity*, not just human activity in general, is impacting the climate in catastrophic ways, yet she continues to take actions that needlessly release carbon into our atmosphere. I'm not judging her. I do the same thing all the time.

In Chicago, a man in his thirties said he feels depressed all the time. He has read about strategies for coping, but the depression itself leeches his motivation to act in a way that could help him feel better.

In Atlanta, a woman told me that she's been cheating on her husband, whom she still loves. She wants to stop, but she just can't break off the affair—she's tried many times to do so. And in Portland, a gay man told me that he lives in constant fear that he'll go to hell for being gay, despite the fact that he hasn't read the Bible in years and has rejected the fundamentalist religion of his childhood.

Of course, long before podcasts explored these questions, and before scientists captured images of living brains in action, the great spiritual and moral teachers wrestled with this same dilemma. People like Moses, Jesus, Muhammad, Buddha, and Richard Dawkins have offered compelling metaphors to help us understand our struggle to live a good life, to be good people in our own eyes. Yet the struggle never stops. Across the entire arc of human history, we've wrestled with the same inescapable question:

Why did I do that?

For most of my life, I was a fundamentalist, Evangelical Christian. I'm not anymore (that story is the subject of my first book). But I am still fascinated by the way ancient wisdom traditions, like Christianity, can bring us solidarity and frame our contemporary struggles as something more lasting.

Adam and Eve eat fruit from a forbidden tree as they listen to a seductive whisper.

Why did I do that?

King David, a man after God's own heart, sees a naked woman bathing on her roof and summons her to his chambers—even after learning she is already married to a soldier in his army.

Why did I do that?

Even Paul of Tarsus, credited with writing more of the Christian Bible than anyone else, has his own take on it. Paul writes, "I do not understand my own actions. For I do not do what I want, but I do the very thing I hate."

Why did I do that?

We hear it from every pulpit and every Bible study. But we also hear it in every weight-loss plan, self-help book, and recovery meeting. People spend billions of dollars per year and countless hours struggling with the same question.

Why did I do that?

The ancient scriptures attempt to describe a life of virtue, but they are short on practical instructions. How do we "think on things that are pure" when our phone never stops dinging? How can we live with peace and gratitude when our lives are measured paycheck to paycheck? How, exactly, does one "renew their mind daily," Paul?

And have modern attempts at self-control faired any better? There's overwhelming evidence that diets don't work—any of them. Self-help books that offer a REVO-LUTIONARY INSIGHT never seem to actually revolutionize the world. In fact, only a tiny fraction of any such book's readers find themselves changed in any meaningful way.

Today, our eyes are starting to open. Neuroscience, behavioral economics, and cognitive psychology offer powerful insights into the question that plagued Paul and the everyday anxieties that plague our entire species. For the first time, we're beginning to learn why our actions are so often out of sync with our intent.

There's more going on in "you" then "you" are aware of. Cues in your environment; the actions of those around you; the words you hear, read, and think; your past experiences; your body's systems; the structures in your brain; and, yes, your own willpower—all of these factors come together to compose your deepest thoughts and feelings, and ultimately what you do.

You've been taught to think of yourself as an individual, but we're learning that's not the most accurate way to view a person. You are a vast collection of brain cells and bacteria, a teeming ocean of life checking Instagram for likes. Though you may think of yourself as the conductor of the orchestra that is your life, you are more like the orchestra itself, a myriad of musicians coming together to create a single symphony.

EACH DAY, I am surprised that my struggle to learn who I am has grown into podcasts that reach an audience of millions. I am not a scientist. I am not a pastor. I don't have any formal scientific or theological education. In fact, I dropped out of community college after six weeks.

The success of my work, launching me from a tech nerd in the advertising business to a D-list Internet celebrity, is a mystery I've been trying to understand for years. All of this started when I began telling people a story about my life: what it was like to be a deeply religious person who stopped believing in God. Every community has taboos that mark some topics as unspeakable, and the religious fundamentalism of my youth was no different.

I started to tell my story because I suspected that many people felt suffocated, alienated, and afraid under the weight of those taboos, and I was right. In the process of defying taboo and being honest about my life experiences, I've met thousands of people around the world, and gotten messages from tens of thousands more. About 12 percent of my audience call themselves Evangelical Christians. Oddly enough, about 12 percent also identify as atheist. The largest bloc is made up of people who don't use any religious or spiritual label at all. That kind of spiritual diversity isn't common in media. How on earth do people with such different perspectives share space?

The common ingredient in this spiritually diverse group is rejection. Almost everyone who listens regularly has been ostracized from a community for questioning, or *deconstructing*, the assumptions of that community. Plenty of them are former Evangelicals, but many are also secularists who had a transcendent experience and found that their unbelieving friends or families didn't want to hear about it. They find themselves living as exiles, grieving the loss of a community that was dear to them. But once they enter the wilderness, their deconstruction process spreads. They begin to deconstruct their assumptions about politics, sexuality, gender, race, and every other component of their former worldview.

Glance at social media, and you will see that this process is playing out across human societies. For many reasons, our species is in the middle of a renegotiation of what it means to be a good person. Vital work is happening to confront the widespread racism, sexism, homophobia, and ableism in our world. I support and celebrate that process. But, as we tear apart all the old scripts, millions of people have shown that they feel lost and confused.

Everyone wants to be a good person, but few have any idea what that looks like twenty years into a new millennium. Many of the cherished lessons from our childhood—we're all the same, we're all equal, Western society is a beacon of hope in the world—have turned out to be fanciful at best. A more complete understanding of history shows us that so many of us who feel like "good people" play an active role in the oppression of people all over the world, including in our home countries. We may believe that everyone is equal, but the data are clear that we're doing a terrible job of treating everyone equally. Many of us have learned how to question and reject harmful assumptions, but we're struggling to create new beliefs and behaviors to replace the older ones we've rejected.

This is a time for action and for serious conversations.

But many of us were raised in cultures that value comfort and civility over truth telling, and because of that we tend to shut down at the very moment we need to rise up.

What's next for us? How do we create the kind of world we want to live in, instead of coasting along as the old one burns itself down?

THIS ISN'T A self-help book. I don't have ONE WEIRD TRICK to help you lose weight. There will be no aha moment that enables you to change the things you don't like about yourself, or eliminate bad habits in just thirty days. That's not to say personal change is impossible—quite the contrary. There have been periods of my life when I've had remarkable success changing habits that I knew were unhealthy.

Twenty-three years ago, I put a gun in my mouth and pulled the trigger. Fifteen years ago, I had such problems with hoarding and organization that I was afraid to let people see my office, my car, or my home. Less than ten years ago, I weighed more than three hundred pounds. I couldn't see my feet when I looked down, and my shoelaces were always tied on the sides because I couldn't reach my feet without crossing my legs.

I have wrestled with my demons, and each time I've triumphed over them, there's been a pattern. First, I'd discover something in the sciences that helped me understand myself better. Next, I would study the work of experts in their fields (and sometimes spiritual teachers) to devise a strategy to modify how I thought, felt, and acted. Finally, I would track these changes over time to make sure that it worked.

Using this method, I've done some amazing things.

I lost more than a hundred pounds and ran a marathon.

My home and office are clean and organized, and I have learned to let go of material things.

My life is full of deep, meaningful relationships that I engage in vulnerably.

And I never, ever, put the barrel of a gun in my mouth.

But this book isn't about those things. In fact, the last eighteen months of my life have looked less like a string of successes and more like a parade of failure, loss, and confusion. I've had mental breakdowns and major relationship conflicts. I have watched one of my children struggle with an eating disorder, been diagnosed with a significant disability, lost a dear friend, and been hospitalized with heart disease.

My successes and failures all have something in common. Each of them has been a chapter in a much larger project: a multi-decade journey to arrive at something far more precious, but also much harder to measure and market.

I like me.

I don't just tolerate me, or accept me. I like me. In fact, I *love* me. I am a huge fan of me, from the way I mispronounce words to the hair on my back.

When I look in the mirror, I see a miracle billions of years in the making, a collection of cells that follow an unbroken line to the very first life on this planet. Even in my most challenging moments, if I plumb the depths of what science reveals about my mind and body, it can lead me to a place of remarkable peace with myself.

Here's the trick: There is no trick. There is no quick, easy road to self-acceptance and grace. I've spent the last twenty years furiously studying the sciences to better understand humanity—not because I was curious, but because I was dying. The stakes for me were life and death, and I am alive.

This is a book about you learning that you are a miracle too. I want to start you on a journey that ends with you looking in the mirror one day, unable to hold back tears, because instead of seeing someone who isn't tall, thin, young, or attractive enough, you instead see a profound and rare being who is worthy of love. I want you to see yourself and be awed, because you are truly awesome.

And I mean awesome in the cosmic sense, not the cultural one.

I want to introduce you to the marvelous miracle you meet in the mirror every morning. I want to show you the unbelievable systems that create your every moment, because it's the key to coping with the times when you feel less like a miracle and more like an unbelievable pain in your own ass.

The very things that make our behavior so frustrating are the mark of a feat that only life can produce: consciousness. Consciousness is a trickster, but it's a remarkable magician. It takes such a diverse cast of characters—the cells in your body, the bacteria in your belly, the light striking your eyes, the pressure waves echoing down your ear canal—and it produces the most amazing trick I've seen in our world. Experiences.

Your consciousness works this magic even as more ancient parts of your body operate with a nostalgia for a world that passed when humans first crafted civilization. Your consciousness is often on the losing side of a never-ending tug-of-war against your impulses, your emotions, and your environment. It takes all this beautiful chaos and weaves it into a single story that you live, becoming aware of it only in the moment when you tell it.

In the rest of this book, we'll look into what science has learned about your brain and your body, and how the two relate. We'll explore the thriving biosphere within you, and how it impacts how you feel. We'll plumb the depths of the best contemporary understanding of the invisible rails that guide your thoughts, and the way the pain of your past remains present with you today. On that journey we'll cover some familiar ground, like cognitive behavioral therapy, or the triune model of the brain. But we'll also explore some less familiar ground, like Accelerated Experiential Dynamic Psychotherapy, supernormal stimulus, and polyvagal theory.

We'll map out why so much of the modern world leaves us stressed out, lonely, and confused, and how words shape our reality more powerfully than we can imagine. We'll learn about the evolutionary role our feelings play, even the ones we don't like. We'll soar through the rarefied air of human relationships and what makes them healthy.

As we do so, you may learn to love yourself a little

more, to embrace feelings you don't enjoy, and be more conscious of the ways you operate on autopilot. You may learn to appreciate sadness or anger, or have an "aha!" about a relationship that once went wrong. You could even pick up strategies for changing your behavior and your thoughts and feelings over time, with a heavy dose of patience, tears, self-reflection, and the occasional bout of excruciating work.

But most of all, I hope you leave this journey with a new appreciation for my favorite thing of all in this universe we inhabit: the miracle of you.

2

Your Brain Is a Burrito

*The Structure of Your Brain,
and Why It Matters*

Sometimes I say things that are so offensive and hurtful, I can't believe myself. The worst part is that I never do it on purpose.

Eight or nine years ago, I was talking with a friend about some problems I had at work. I had botched a performance review earlier that day, and it left me feeling frustrated with my competence as a manager. After telling the story, I said, "It makes me want to kill myself."

That's an odd phrase to throw around casually—especially for a suicide survivor. More important, and the real reason that my words shocked me so much, was that the friend I was talking to had recently lost his son to suicide.

In what seemed like slow motion, I watched an expression of pain cross his face. He collected himself quickly, and then put a hand on my shoulder. I'll never forget the warmth in his eyes as he said, "I bet you do, but remember, your life is valuable." And then he walked away.

I felt like a heel. Even today, my cheeks burn with shame as I write that story. But there have been other times when I've said things just as embarrassing and accidentally callous, like the time I described a parking regulation as "retarded" to my friend Lisa, whose daughter has Down syndrome.

I want you to know that I am a sensitive person, and that I try to be careful with my speech. We'll talk more about this later, but my childhood was an extended lesson in how bullying can shape a child's identity. I care a great deal about using words that help and heal instead of harm. But I still made an offhand remark about suicide to a grieving father. I still called something "retarded," even though I abhor that word and the way it's been used to mock, erase, and minimize people with disabilities. Both incidents make me angry when I think about them. How could I have been so thoughtless?

Isn't it strange that we have the capacity to say things we don't mean to say? In the worst cases, moments like these can raise questions about identity and self-knowledge that last for years.

Language isn't the only thing we struggle to control. Other times our behavior runs amok in ways that are no less frustrating, but at least are easier to rationalize. I can say I want to exercise, for example, but to do so requires effort. I have to put on workout clothes, set aside a half hour, and exert my body to the point of discomfort. But I begin with out-of-control language because it's the crux of the issue I want to explore in this chapter: the biology behind why we so often can't control ourselves, and why it feels so confusing. How on earth can I say things that I don't mean to say?

To answer that question, we've got to start at the beginning—of life itself.

YOUR ANCESTORS WERE brainless. Don't worry, this isn't the beginning of a roast. I mean no offense here—you and I have the same ancestors, and we share those ancestors with dogs, crocodiles, dinosaurs, potatoes, palm trees, and the bacteria that gave you a sinus infection last spring. Your brain, in particular, is billions of years in the making. I mean, think about it: It took more than three billion years for life on earth to produce something that could make and use Google Maps. As far as we know, there's nothing like your brain in the entire universe. Here's how your incredible brain came to be.

We begin 4.54 billion years ago, when the earth first formed. It was a desolate hellscape of molten rock under a nonstop barrage of rocks that smashed into it from space. Many of those rocks brought water in the form of ice, which became steam on our primitive, superheated earth. Some 4.41 billion years ago, the earth cooled enough for the steam to form oceans and create a remarkable mix of water and organic material.

We're not exactly sure when life formed in that primordial stew, but we know it was between 4.28 and 3.77 billion years ago. Our earliest ancestors—the common ancestor of all life on this planet—were microbes that fed on organic material spewing from hydrothermal vents (basically, holes in the seafloor where magma meets ocean water and creates a plume of hot, nutrient-rich water). They had no brains. How could they? They were single-celled, without nerves of any kind. They had no need for them—the whole point of a nerve is to carry information between cells in a multicellular organism.

Thanks to evolution, these ancestors diversified into many types of single-celled organisms. Some still "ate" from thermal vents, but others "ate" other organisms, creating the first food chains. Around that time, some cells started clumping together after dividing, which helped them avoid predation. The first multicellular organisms faced unique challenges of their own. A critical one was the need to organize actions across their many cells. So, while sponges, for example, lack nerves of any kind, they can use "calcium waves" to "sneeze" out contaminants and keep themselves healthy. Other primitive animals can organize electrical signals in a specialized form of cell in their digestive systems.

A little more than 740 million years ago, a remarkable innovation appeared in the animal kingdom—jellyfish. These early animals were the first to have a nervous system. Jellyfish have nerves spread evenly throughout their body in a netlike structure. They have nothing that resembles a brain, but their nerve nets allow them to feel when they're touched, to detect chemical signals, and to coordinate actions. Because of this, jellyfish can swim, catch prey, and respond to attack.

The next major innovation came from flatworms, around 550 million years ago. Their nerves converged into a central cord running from front to back, with a large cluster of nerves at the top of this cord. This bundle is called a "ganglia." (I giggle almost every time I hear the word "ganglia," because it sounds dirty somehow.) Ganglia in flatworms were the first primitive brains to appear on earth. And I do mean primitive: Worm brains consist of dozens to a few hundred neurons, while you and I have about 86 billion.

Brains radically increase the capacity for animals to search for food, avoid being eaten, and find mates. (Sex was a relatively new idea 550 million years ago.) So, the emergence of brains led to a massive surge of diversity in body structures, which we call the Cambrian explosion. Most of this new life followed the flatworm's nervous system structure, with a nerve cord and a brain.

From there, branches continued to sprout on the tree of life, and many of them helped pave the way for your brain. Arthropods (crustaceans, insects, and other animals commonly called "bugs") offered a major innovation over their wormy predecessors: doubled-up ganglia arranged in bilateral symmetry. By the time the earliest fish appeared some 360 million years ago, their brains bore a remarkable resemblance to the structures we find deep in the brains of modern humans. These sophisticated brains allowed the formation of complex eyes, and fine motor control for coordinated swimming. More important to our story, however, the complex brains of early vertebrates allowed them, and now us, to experience something remarkable: emotions. Anything with nerves can have physical sensations of pain, touch, and hunger. But the large brains of vertebrates offer a richer tapestry where sensations and stimulus converge into feelings.

Amphibians introduced remarkably intricate brains millions of years later, with distinct structures that coordinated their behaviors and survival functions. The brain of a frog controls its heart and breathing, all while building a model of the world that frogs can use to navigate a challenging world.

The earliest mammals emerged some 225 million years ago. To describe why that matters to you today, we need to take a moment to look at how evolution handled brain development so far. Jellyfish had the first nerves, but no central nervous center. Their nerve nets developed into a nerve cord with a simple ganglia in the "front" of a flatworm. This ganglia grew into distinct left- and right-side structures in arthropods. Then, for the first vertebrates, the nerve cord was encased in bone, and the ganglia were surrounded by complex structures that became a true brain. In essence, evolution keeps piling new structures around older ones—without ever really replacing them. We'll get into specifics later in this chapter, but keep this general idea in mind: Evolution keeps building new brains on top of the older ones.

Our mammalian ancestors still had the ancient amphibian brain, which handled respiration, heartbeat, and other involuntary body systems. But they also developed newer structures to handle socialization and executive function. Mammalian brains were much larger, and offered our first warm-blooded ancestors the ability to experience powerful emotions: love, anger, and fear. Emotions create strong motivations to survive and protect one's kind, and they helped mammals succeed as a class (especially when the dinosaurs conveniently went extinct).

When early mammals left the forest floor for the trees some 80 million years ago, their brains grew yet another feature: an intricate outer layer of brain tissue that was densely packed with neurons. And, by the time these early tree mammals developed into the first hominids 4 million years ago, their brains started to resemble what we have today. Hominids used those superpowered brains for making tools and, starting roughly 100,000 years ago, communicating with language.

Here's the point I want you to remember: Brains were around for hundreds of millions of years before the invention of language. Which means language is a new, relatively unproven strategy for organisms trying to survive and reproduce.

I live in a house built in 1946. It's got a rather strange layout, because families that lived here before mine added rooms and renovated as they needed them. Just as no architect would sit down and design my house's layout from scratch, no engineer would design a brain like yours. But evolution never gets to start with a clean slate—her designs have to survive and reproduce, so she's in a continual state of testing them on the fly. This piecemeal manner in which your brain evolved ended up creating the astounding and incredible thing we know as you.

• • •

YOUR BRAIN WEIGHS about three pounds, and is roughly the consistency of tofu. If you removed it from your skull and set it on a table (and I don't recommend that you do), it would sag under its own weight. Because your brain is so important to your survival, millions of years of evolution have surrounded it in an armor-clad enclave of bone, and suspended it in fluid so it's comfy in there.

There are something like 86 billion neurons in your brain, along with another type of support cell, called glial cells, which also number in the billions. Those neurons are connected to each other in networks that crisscross your brain like a three-dimensional road map much more complex than the highways of major population centers. To communicate with each other, neurons zap each other with electricity and exchange chemicals. Somehow, all of that becomes you. When you laugh at a movie, weep with a friend, get hungry, fall in love, or remember a night out with your friends, all of it is happening in three pounds of tissue the consistency of tofu, zapping itself with electricity while secreting and absorbing chemicals.

Your brain is by far your most complex organ. I don't mean that as an insult to the remarkable branching network of tubes and air sacs in your lungs, or the miraculously coordinated chambers and valves of your heart. But while your lungs' key function is to exchange oxygen for carbon dioxide, and your heart's is to pump blood, your brain . . . well, your brain does a lot. Your brain isn't just an organ that "thinks." It's the organ that ties all the other systems of your body together.

In fact, I believe some of the problems we have understanding ourselves come down to the misguided notion that we have "a brain," instead of a massive, sprawling set of brain structures and neural networks competing to influence every moment of our lives. All of those structures emerged at some point in evolutionary history, and they survived because they can solve a problem your ancestors faced. You get motion sickness because your brain evolved to detect new poisons in plants based on how they impact your senses. You get cold hands before giving a presentation because your ancestors needed to minimize bleeding during altercations with other animals.

I find it helpful to visualize how your brain is structured, because that structure can help you understand your behaviors—especially when you're stressed. Let's start by thinking about a burrito. I'm going to think about a carnitas burrito, because that's what I ordered yesterday in Los Angeles, when I got to watch the formation of a truly delicious burrito for my lunch.

First, the deli staff prepares the carnitas. It's the foundation for everything to come. Juicy, tender pork that's been marinating for hours is pulled and set aside on a flour tortilla. Next, I watched as refried beans, rice, and some kind of sauce were added, and the tortilla wrapped tightly around the whole concoction. Viewing this culinary marvel, I thought the only reasonable thing one can think when handed a hot, fresh burrito: "Wow, burritos are a lot like brains."

I'm serious. A great burrito starts with some meat and refried beans, just as your brain started with a set of ancient structures that are often called the basal ganglia, brain stem, or "reptilian brain." Why the reptilian brain? Because, like we just saw, this part of your brain looks very similar to the brains of early fish, amphibians, and reptiles. These structures manage all your involuntary functions (like your heartbeat and breathing), as well as ritualistic behaviors and territoriality.

Your basal ganglia are running the show when you defend your place in line at a concert from "cutters," or when you get lost in thought while driving on the highway. Try not to do that, by the way. Your basal ganglia lack the strategic capacity to deal with unexpected developments on the roadway—and may be more comfortable rear-ending another car than are other parts of your brain.

As our burrito brain grows, spoonfuls of rice, sauces, and other ingredients get piled around the fundamental meat and refried beans. In your brain, this next layer of structures is often called the limbic system, or "old mammal" (paleomammalian) brain. These are the parts that coordinate your social behaviors, nurturing, and reciprocity.

Finally, all these ingredients are wrapped inside a tortilla—that's the neocortex, or neomammalian brain. Interestingly, the neocortex is roughly the same thickness as an actual tortilla, although your brain's tortilla is all wrinkled up, increasing its apparent thickness. Your neocortex does all the fancy, human things: language, advanced cognition (like planning for the future or modeling reality), music, and advanced visual processing.

Among neuroscientists, this "burrito" view of the brain is called the *triune brain model*. It's an older way of looking at the brain, and full of faults, mainly because it ignores how remarkable and evolved the brains of modern birds, reptiles, and fish are. But for our needs—understanding your brain is not "one" thing—the triune model is a helpful place to start. I often think of my brain as a person standing on a dog, who in turn is standing on a crocodile. Most of the time, they get along pretty well. But things get dicey when something unexpected happens and you're forced to make a quick decision.

If someone cuts me off in traffic, my brain interprets the unexpected intrusion as a serious threat. The crocodile gets first shot at anything like that, and it sends an unambiguous signal to the threat: a car horn and raised middle finger. (If crocodiles had middle fingers, you know they'd use them.) The crocodile's first question in any situation is "Am I safe?," which is convenient for something located so close to the brain stem and spinal cord.

Let's say I'm standing in line to put my name in at a trendy brunch place, and someone gets in line in front of me. The crocodile has a rapid, decisive response to territorial intrusions—that the best defense is a good offense. But the dog, who springs into action next, sees the situation differently. Before acting, the dog (or limbic system) wants to know if the person in line is someone we know or not. Is this guy a member of the pack? If so, wagging their tail or offering a warm greeting would be a better response than lunging at the offender with teeth bared and eyes closed. The dog is a master at reading body language and nonverbal communication, and can rapidly assess what others are feeling. Their first question in any situation is

"Do I belong?" They're concerned with feeling loved and creating a social standing with others—and they know the crocodile can be too aggressive at times.

The human in my head is often lost in thought and slow to respond. I picture him reading a newspaper all the time, only looking up whenever there is a commotion. The crocodile and the dog have already snapped into action. But when the human (or neocortex) catches on, his concerns are entirely different. He knows it's important to take time to hear what others are saying—maybe he'll find that the intruder is meeting friends, or just checking how many people are on the list in front of him. The human knows there's an elaborate set of laws and social conventions to navigate in situations like these. The neocortex is much slower to act than the reptilian brain or the limbic system, but given time to process, it can do a much better job navigating the world we live in.

This dynamic plays out dozens or more times in our lives every day. Our senses send stimuli to our brains for processing, but once they arrive, they get handled by brain structures that have different, and often competing, agendas. The crocodile, the dog, and the human have different priorities.

This analogy is drastically oversimplified. Your brain isn't made of three structures; it's made of hundreds.* My point here is that the brain stem, limbic system, and neocortex are subdivided into specialized structures, all of

---

* For a more detailed profile of the most prominent players, organized by what part of your brain they reside in, check out Appendix A toward the back of the book.

them fighting to have a say in how each moment unfolds. Sometimes these structures work together. For example, your amygdala and your orbitofrontal cortex have a long-distance relationship that spans several layers of brain tissue. Though located far apart, they "talk" to each other often—a fact that leads to your capacity to experience anxiety. This makes *so much sense*. The part of your brain that anticipates the future needs to be able to motivate you to action. It does that by forming a network with your amygdala and coordinating to produce a feeling more sustainable than fear. We don't enjoy anxiety. We don't like it. But we're not supposed to like it. Anxiety is the brain's way of trying to make you do something about the stressors in your environment.

Your brain has hundreds of documented networks like this. If you believe in God, you've got a "God network," and the structures involved affect both your theology and your experiences with religion. You've got networks for dealing with surprises, for mapping out social networks, and for creating complex emotions.

The dissonance between these networks creates the inner conflicts we experience every day. One network derives pleasure from the bright lights and creative storytelling available on Netflix, while another network understands tomorrow will be a rough day if we don't go to sleep soon. One network generates pleasure when we bite into a truffle-flavored potato chip, even as another network says our stomach is full, and a third forecasts what number the scale will display when we step on it tomorrow. Meanwhile, another network tasked with understanding social rank processes that weight forecast and fears you could suffer socially if you gain weight.

You are a miracle because 86 billion neurons in your brain form into thousands of structures and networks, built from a map created over billions of years to understand the world you live in. But sometimes, you are a pain in your ass because all these networks are running a playbook that's been around a lot longer than you have. The cells in your body have survived through the eons by eating every delicious calorie they come across, allowing fear to make them run, and using anger to make them fight for their lives.

Your brain isn't here to show you an honest picture of the world, or to make you happy. It's not here to make rational decisions, or to help you advance up the corporate ladder. Your body spent the energy it takes to build something as powerful as your brain for one reason and one reason only: Your brain helps you survive on a planet that is often hostile to life.

EVERY PERSON IS different, and every human brain is different. When we look at brains through the lens of neuroscience, we're mapping out the features that we see in common across entire populations of people. But really, this work is only just starting. For example, we don't have enough imaging studies with samples of left-handed people to understand the relationship between the two halves of the brain.

We start out with a genetic template, and it interacts with our environment to create a completely unique human being. This variety in brains, and in how they resolve conflicts between networks, means that some people are more prone to overeat, and some more prone to exercise. Some people get angry more easily, while others are more naturally sanguine. And this makes sense as an evolutionary strategy. Some brains are more capable of survival in famine, and others in times of plenty. As weather patterns change, civilizations rise and fall, and cultural norms change, the diversity in human behaviors helps ensure our species sticks around and adapts.

Honestly, though, I've been frustrated with my brain my entire life. Some people are surprised when I say that— many of them listen to my podcasts and see me as a successful creator and communicator. In many ways, they aren't wrong. I'm good at absorbing and synthesizing information, and I can be an emotive and compelling storyteller. But I haven't always been that way.

When I was a kid, I was different. I had trouble reading and writing, and difficulty making friends. Teachers often complained to my parents that I was lazy, but the instructions they gave me in class confused and frustrated me. The act of forming letters with a pencil was so difficult, it brought me to tears.

Crying at your desk is an easy way to get beat up during recess, so I learned to shove down my tears and escape into my imagination. It helped me cope with the fear I felt toward my classmates, but it made me look obstinate to my teachers. I didn't know how to communicate my fundamental struggle with doing what they asked in assignments—I was just a little kid.

I got Cs and Ds all the way into high school. Almost every year, there were serious talks about holding me back for a year—and in first grade, I spent time in the "special" class with one-on-one instruction. Over time, though, I learned to compensate for my struggles. For example, I still have trouble filling out a form by hand, but typing on a keyboard is basically effortless. I have trouble understanding people when they talk, but if I look at their lips and imagine I'm close captioning what they say, not only can I understand them, but I can also remember much of what they say.

I've got a million little strategies like these to get me through the day. They allow me to look normal, even successful, to other people. But there are some things that I've never been able to compensate for, and shame has taught me to hide those things from others. For example, I often spin in little circles when I'm alone. When I do this, I make a robot noise in my throat. If I'm stressed or anxious, I can do this for as long as forty-five minutes or more.

I like to eat the same thing every day: a vegan smoothie for breakfast and then chickpea salad for lunch. At dinner, I strongly prefer to eat one item at a time, in a clockwise pattern around the plate. Similarly, I have a ritual for getting ready each day, and I get confused and disoriented if anyone interrupts it. My wife realized years ago that if she asks me a question while I'm brushing my teeth, I'll almost certainly lock myself out of my car, office, or home later that day.

I'm dependent on my routines to function. But none of that makes me feel as much shame as this: If I have an appointment on my calendar, and the other party misses it—or worse, if I wrote it down wrong when the appointment was made—I completely and utterly lose control of myself.

It begins like a panic attack. My stomach tightens as a tide of fear and anger begins in my belly and rises like the storm surge of a major hurricane. As that wave rushes over me, I will often begin to scream, sometimes wordlessly, and other times by unleashing the most self-annihilating stream of obscenity you've ever heard. I've sat in my car and pounded the steering wheel until my hands ached. I've screamed until I lost my voice. All of that because I wrote "10" instead of "12," or someone was double-booked.

For years, I understood these behaviors were abnormal, but I never questioned why I did them—let alone asked a doctor, or a therapist, or any medical professional for an opinion. I just assumed they were a shameful yet unchangeable part of me. They caused me frustration and grief, but I was able to hide them well enough. At least, until I started working from home.

JENNY AND I both stay home full-time. I have meetings around town, and I go to the Liturgists office for a few hours a couple of days per week. But in recent years, Jenny and I have spent more time together than ever before. That's made it hard for me to hide my "quirks."

If I'm researching an answer to a question for my podcast *Ask Science Mike*, and she walks in and interrupts me, sometimes I'll snap my jaw. But if I'm in a particularly deep state of concentration, I may grab a handful of my hair, or even slap my face. I'm only vaguely aware of this when it happens, but the expression on Jenny's face helps me realize what I'm doing.

Jenny has often remarked that she believes I'm on the autism spectrum. It began early in our marriage, as a way of defusing the tension that bubbled up when I was acting particularly rigid. Over time, those quips evolved into more serious suggestions.

I sympathized with her, but didn't buy it. I told her that any resemblance I had to someone on the spectrum was due to media tropes, not medical criteria. She'd drop it for a few months, and then mention it again on one of my bad days. I didn't take her seriously because Jenny isn't trained to diagnose autism spectrum disorder.

But Jenny wasn't the only one. People email me thousands of questions a month—that's what my podcast is about. They also stand up and ask me questions at live events. A couple of years ago, listeners with autism started asking me how I cope with being an autistic person and a public figure. I was never sure what to say. I felt honored that people trusted me enough to open up, but I also felt terrible because I didn't know how to answer. How could I understand what life was like for a person with autism?

I'd respond by thanking them for their question and telling them that I was sorry they had misunderstood me, but that I wasn't autistic. Many times, a look of disappointment would form on their faces as they sat back down.

But more than a decade after Jenny first suggested I could be on the spectrum, I was talking with my friends Michael Gungor and Hillary McBride in Austin, Texas. Michael and I cofounded The Liturgists together, and Hillary is one of the cohosts of *The Liturgists Podcast,* and a licensed psychotherapist who will soon complete her Ph.D. She knows her stuff. We were in town for The Liturgists Gathering—an event that gives the misfits who listen to our podcast an excuse to be in a room together.

It was the kind of conversation that involved tacos, a pitcher of margaritas, and a lot of giggling. As we passed the halfway point in our pitcher and traded jokes with bellies full of tacos, Hillary did a shimmy in her seat. After Hillary shimmied, Michael did as well, and then they both started laughing. Now it was my turn.

I don't understand how people move their bodies to dance. I have very limited control over the muscles in my torso. Everything from my shoulders to my hips feels like a single, fixed structure. So, when I tried to shimmy, I looked more like a lawn sprinkler than a dancer. It was hilarious in its own way, but not the desired effect.

Hillary noted my frustration, and later that night as we all unwound over drinks in our Airbnb, she asked if we could talk. "Mike," she said, "I haven't known how to bring this up, or even if I should, but have you ever considered that you could be on the autism spectrum?"

I told her Jenny had said this too, and so had some of our listeners, but I knew I wasn't autistic for two reasons. "I don't have trouble fitting in during social events, and I look people in the eye when I talk to them," I said.

Hillary nodded. "How do you feel when you look people in the eye?"

Without thinking, I replied, "Horrible . . . but I just repeat, 'LOOK INTO THEIR EYES,' over and over in my head. I've learned that people don't feel seen or valued if you look away while they are speaking."

Hillary said, "You aren't unconvincing me."

Then, because Hillary is a professional, she made clear that I was not her client, and she was not diagnosing me, but merely having a conversation between friends. Still, Hillary's thoughts carried weight, given her experience and qualifications. The following week, I took the more serious turn of searching the Internet for autism screening tests.

I took more than thirty-five that night, and all of them told me I should consider a formal test. So I did—the very next week.

LAST JUNE, I met with a psychologist to see if the suspicions of those closest to me had any clinical backing. Autism testing for adults is uncommon, so I found myself in a testing center meant for children.

After introducing herself, the psychologist asked what brought me in, so I told her. She repeated what the receptionist told me, that this center only worked with kids, but she said she had a few hours open that morning.

So I took some screening tests. She interviewed me, asking lots of questions about my childhood. But something about the interview bothered me: She kept asking about things I'm embarrassed about, and that I hide from other people.

She asked if either now, as a child, or both, I've experienced temper tantrums that I couldn't control. When she did, my eyes filled with tears, and I thought about all the moments when I've sat in my car and screamed.

She asked if I ever liked to arrange my toys in rows or lines as a child—which was one of my favorite things to do when I played alone. She also asked about my current life, and what kinds of routines I had. So, I told her about the patterns of behavior I mentioned above, along with other observations like how I get ready in the morning, and how many times I turn to the left every time I shower.

She asked if I ever hum when I'm by myself. I said no. But then she pressed, "What about other noises with your throat?" I wondered if she had a microphone hidden in my car or in my office.

"Mike, do you have any trouble sitting still?" she asked. I told her that I didn't, so she asked me to sit as still as I could and look her in the eyes. I did so, and thought how easy a test this was, until she said, "What's going on with your toes there?" I looked down and saw my toes wiggling in my shoes, something they do often enough that they tear open my socks and carve holes in my house slippers.

"Oh, that happens all the time," I said. "I don't really notice—"

She interrupted me. "Can you keep looking at my eyes while you speak?" *Of course,* I thought, and then looked in her eyes while trying to speak. I couldn't. My eyes darted away for a moment, almost involuntarily, as I quickly put together a few sentences. "Of course I can look into your eyes and speak. That's a really elementary facet of social integration."

The psychologist asked a few more questions, and then left the room to review my tests with a colleague. She didn't have any magazines in her office that weren't meant for children, so I took some time to catch up on Twitter.

When she returned, she went over the results with me. Based on her screening tests, I wasn't likely to have obsessive-compulsive disorder or social anxiety disorder. I was significantly face-blind—which explained why she'd given me a test that involved trying to match pictures with faces (more on that later in this chapter). Then, she got to the reason I'd come.

"Mike, autism is notoriously difficult to diagnosis in adults. The *DSM-V* leans heavily on childhood behaviors for good reasons: Adults who can make it to adulthood without a diagnosis are very good at hiding most of their symptoms. There aren't many clinicians who specialize in diagnosing adults, even here in L.A., and I want you to know that diagnosis requires people from multiple disciplines weighing in. It requires looking at records from pre-school and early elementary school. If possible, it involves interviewing your parents. Diagnosis is much, much harder for adults who weren't diagnosed as children."

I took a deep breath. I'd read most of that on the Internet, but hearing it from a real professional, in person, was much more powerful.

"I can't diagnose you with autism spectrum disorder. But I can tell you, based on these tests and what you've told me about your childhood, you're certainly a candidate to be on the spectrum. I've got a friend who specializes in adult diagnosis and treatment. Would you like me to contact her and see if she'll see you?"

I said yes. A few days later, I got an email from her friend, which started a longer, more arduous process of evaluation that ended in hearing this sentence: "Mike, you fit the diagnostic criteria for autism spectrum disorder."

IT'S SHOCKING TO learn you have autism when you are forty years old. It's even more shocking if you are in the process of writing a book about self-knowledge and self-acceptance. I felt insecure. How could I have missed something so significant about myself, and for so long?

But it didn't take long before I felt relieved. So many of my eccentricities and struggles can be described with a single word: autism. I actually find that really comforting. I'm not broken, I'm autistic. Learning that has helped me understand myself, and to accept the parts of my brain that I'd hidden for years. That acceptance has helped me feel less shame, less anxiety—and yes, make positive changes in my life.

That's why I told you about learning I'm autistic. You may or may not be autistic, but I do know that your brain is as unique as my own. The template laid out in your

DNA has been shaped by a series of life experiences that belong only to you. There is no other brain like yours, and that means your process of self-acceptance and growth must be different from mine as well.

You may struggle with porn more than I do, but less than I do with pizza. Maybe it's easy for you to exercise, or maybe you're more prone to depression than I am. But, however you are wired, and however that wiring has been reinforced by your life, I want you to know something:

You aren't broken either.

You are here. You are alive. You have survived the challenges of life, and each part of you—even the parts you don't like or appreciate—has played a role in your continued existence. There isn't a piece of you that hasn't been fearfully and wonderfully made, tested by the unforgiving process of natural selection.

Here's what I mean. There's a data-driven model for testing personality called the Big Five. Most personality models have serious issues with reproducibility and consistency, but the Big Five model is based on data instead of theory. It measures personality on five spectrums: openness to experience, agreeableness, extraversion, conscientiousness, and negative emotionality.

I asked Jenny to take it, and she scored a 92 out of 100 on negative emotionality—which is a measure of how prone you are to anxiety, depression, and emotional volatility. That's a relatively high score (Big Five scores are based on a percentile placement, not an absolute scale), so I wanted to learn more about the research. I found out that negative emotionality is a good thing in many ways.

People with higher scores are more suspicious and skeptical—meaning they are less prone to be manipulated or deceived. This immediately rang true for me. Jenny is always on the lookout for potential dangers, often seeing hazards in new social relationships that I've completely missed. Jenny's stable, heritable trait of negative emotionality turns out to be essential for the functioning of a family or group of friends.

After reading about negative emotionality, I walked out of my office and told Jenny that she's a guardian. Her reaction was classic Jenny: "I don't need science to tell me that. You'd all lose your heads without me."

We have unique brains, and our unique brains need each other's strengths. I mean, people tell me that the hyperfocus and routine dependence common to people with autism are the hallmarks of disability, but as an autistic person, I sometimes think neurotypical people are chaotic and have trouble focusing on important matters. In its wisdom, evolution created a remarkably diverse species in *Homo sapiens*, but our human cultures do a terrible job of acknowledging the value of that diversity.

As I tell you stories about myself and others in this book, understand that I'm not setting up yet another example for you to compare yourself against. Instead, I am trying to help you discover your unique strengths, and how they emerge from the completely unique organism that is you. Our work in this book is to take apart and unpack social assumptions, not to reinforce them. Your burrito brain is a one-of-a-kind creation—just like everyone else's.

3

Pizza and Porn

How Supernormal Stimuli Shape Your Life

What do Tinder and turkeys have in common? They both help us understand how our decisions are often hijacked by instinct, as researched by a zoologist named Niko Tinbergen.

Tinbergen studied birds, fish, and insects all his life, starting with a child's natural curiosity and ultimately making a career out of it. In his research, Tinbergen was fascinated by what he called "spontaneous" behaviors, the actions animals will take without specifically learning them, and which resist any sort of training or conditioning. He studied this so much, and for so long, that he won a Nobel Prize in 1973 for his contributions in helping humanity understand how animals' instincts work with their learned behaviors to help them survive.

I find two particular areas of his work fascinating. The first is something called Tinbergen's hierarchal model, which is a way of understanding why animals do what they do. This model became the basis for my absolute favorite image on Wikipedia, a flow chart describing what happens when a bee or hummingbird feeds:

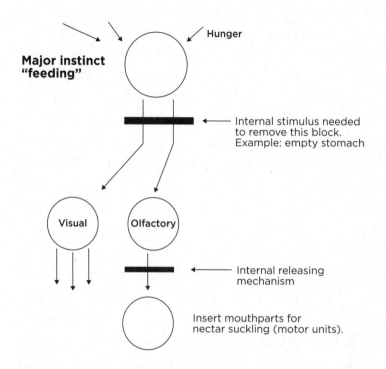

Before we get into why this image (and the model it represents) is so important, let's take a moment to appreciate the most accidentally funny description I've seen in scientific reference materials: "Insert mouthparts for nectar suckling (motor units)." Someone needs to print that on a T-shirt.

Anyway, back to why this model is brilliant. If we start up top, those arrows represent all the features in a hummingbird's body that create a sensation of hunger. You can imagine the topmost circle as a reservoir that "fills up" as different internal systems signal hunger.

But animals don't feed all the time. If they did, they would damage their stomachs, never sleep, never mate, and ultimately die. So, beneath that first circle is a solid black line, representing a "block," or inhibitor. When the first circle fills up, the animal feels hungry, but it won't start feeding unless the block is removed. In this case, an empty stomach is the stimulus that removes the block and allows the next circles to start filling up. The bird "fills up" her visual circle by moving toward a food source—in this case, a flower. At the same time, she will "fill up" her olfactory circle by seeking out the smell of nectar. Once those circles fill, another block is removed, and the bird's "mouthparts" can finally be inserted for suckling.

Dr. Tinbergen imagined this model not only as a metaphor, but as a set of literal structures in the brain. He pictured some kind of neural energy building up in the brain, with gates that opened up new pathways—something we now know isn't true. But though his model is not a good descriptor of brains in a literal sense, it's a great tool for understanding the systems that allow life to exist in a dangerous world.

Tinbergen didn't stop at modeling how creatures respond to stimuli. He started to wonder what would happen if you created a stimulus that was more powerful than what an organism would find in the natural world. How would the animal respond? His experiments on this question offer some shocking insights into animal behavior.

In one scenario, Tinbergen created abnormally large eggs out of plaster and painted them with bright, Day-Glo colors. When he placed those eggs into a nest, he found that a mother bird would try to incubate the fake egg, even at the expense of her real ones. Evolution somehow offered an instinct that said bigger, brighter eggs held potential for bigger, healthier baby birds. And this instinct allowed the mother's "reservoirs" and "blocks" to be overwhelmed.

It wasn't just eggs, either. Tinbergen also found that if you manufactured a fake baby bird with a larger, redder mouth and placed it in a nest with real baby birds, the mother bird would try to feed the imposter at the expense of her real chicks. Her instincts have been forged in a world where food is scarce and not all chicks are equally healthy, so it makes sense that she would favor the chicks most likely to survive. But an artificial stimulus can be introduced that hijacks that instinct in dangerous ways.

Turkeys are even worse. If you craft a particularly comely female turkey head and mount it on a stick, Tinbergen found, male turkeys would try to mate with it—all while ignoring nearby female turkeys. You read that right. When a male turkey's instincts are overwhelmed by an intensity of stimulus that doesn't happen in nature, it will try to *mate with a stick* instead of an *actual female turkey*.

I think that does a good job of explaining how Tinder works.

THINK ABOUT IT this way. Imagine for a second that you're standing on a savannah in southern Africa. Picture some hills in the distance, with the sun setting above them. Nearby, a landscape of ocher soil, shrubs, and dry grasses rustle in the breeze. The world is painted with a palette of earth tones, greens, and the red-orange blaze of a sunset. It's the kind of vista our senses evolved to take in, to search for, and to be inspired by.

Now, imagine Times Square in New York City—bright as day, all night long. A human animal standing in Times Square is bombarded by sights, sounds, and smells that can only be described as supernormal stimuli. Supernormal stimuli surround us all, whether we're standing on Forty-Second Street in Manhattan, New York, or scrolling through Facebook in Manhattan, Kansas. The world that we, as a species, have created for ourselves is nothing like the one we evolved in—and yet we wonder why our behaviors are so hard to control.

Consider one staple of Times Square: pizza. I can't think of any food that's a better example of supernormal stimulus. Nothing found in nature is as salty, fatty, and gorgeously full of carbohydrates as a New York slice. I am obsessed with pizza, and lose all willpower when I'm around it. If there is pizza, I will eat and eat until it is gone, no matter how full I feel.

It's probably killing me. I'm overweight, toward the top of my body's yo-yo cycle as I write this. There's absolutely no doubt that the amount of pizza I eat and the frequency with which I eat it is one of the top three health hazards in my life. I can't help it, though. It smells so good, and I can't look away when I see it. Every time I take a bite of a slice, I feel happy, even joyous, as my brain is flooded with chemicals designed to tell my body "That's good! Do more of that!"

And of course they should! In pre-agrarian times, a pizza would have been a feast of incredible fortune. That many calories, in such a dense presentation, would be worth some discomfort to consume; after all, hunger was no stranger to our species in those days. Is it any wonder that obesity is such a widespread problem when so many of our foods have been optimized, even engineered, to appeal to the deepest, most ancient structures in our brains?

Of course, it's not just pizza and other processed foods. Each day, it seems, there's a new study describing how the smartphone in your pocket is as bright and attention grabbing as any billboard in Times Square. Plus it does the most wonderful thing: that delightful little *buzz buzz* that happens when you get a notification on your "silenced" phone. Sure, your bank could be buzzing you to warn that your account is overdrawn, but it could also be Instagram telling you that someone liked your selfie in front of a well-known landmark. If you're lucky, it could even be a match from a dating app!

The *buzz buzz* of your smartphone represents a potential, but not guaranteed, reward. Studies have shown that these kinds of signals flood our brains with dopamine, the chemical that (among other things) helps create craving in your brain. And when you give in to the signal, a rich, colorful screen with stereo sound immerses your brain with vivid experiences your ancestors couldn't imagine. So, yeah, it's hard to set your smartphone down at dinner.

Supernormal stimulus isn't limited to the objects we create in the world, like pizza or iPhones. We're also performing the same kind of enhancements on ourselves.

Consider the example of Kim Kardashian. Now, before you think I am unnecessarily singling out Ms. Kardashian, hear me out. When I say humans like Kim Kardashian don't exist in nature, I'm not talking about rumors regarding plastic surgery—that's absolutely none of my business, and I don't care. I picked Ms. Kardashian because she's got a significant presence on Instagram, and because her supernormal stimuli strategies, far from being rare, are incredibly common.

Ms. Kardashian uses makeup to change the apparent shape of her eyes and the contours of her face. And makeup is designed, in part, to make women look sexually aroused. Blush on a woman's cheeks simulates the flushing that happens in arousal, while lipstick mimics the enlargement and color shift that happen in a woman's lips. Today, these effects are pushed even further with digital filters on platforms like Instagram, turning the faces of women into nearly cartoonish figures, and reinforcing a cycle where men primarily relate to women as objects of sexual desire.

I'm not saying that women only or even primarily wear makeup to get attention from men. To say that would erase the fact that some women aren't attracted to men, and that plenty of men and nonbinary people wear makeup too. People wear makeup to feel confident, to have fun, and to show off artistic mastery to other people (especially others who wear makeup). But if you think back to our poor, hapless male turkeys clamoring for a moment with a stick, you can see how when one person uses makeup in a way that triggers instinctive behaviors in human males, it sets up an arms race where women who opt to display their natural faces find sexual attention harder to get (if that's something they want at all).

Through the lens of supernormal stimulus, we see women's fashion in a cruel light: a meat market designed to display women's bodies for the pleasure of others. To those who fit cultural ideals, their features are turned into a parody of what is found in nature, while women whose hips, bellies, or breasts are deemed "too large" or "too small" find that every outfit on every rack spotlights their insecurities. Is it any wonder why so many women have eating disorders, hate their bodies, and struggle with a basic sense of self-worth? They are fighting against forces much larger than themselves.

Men are every bit as guilty of this phenomenon. My torso, for example, is more of a pear than a V, but if I throw on a sport coat, my love handles vanish, while the jacket's shoulder pads create an appearance of muscle mass that isn't there. I trim my beard carefully each morning to present an illusion of sharpness in my jawline, and my hair is cut in a way that minimizes the prominence of my thinning mane. I am every bit as much a practitioner of supernormal stimuli as Ms. Kardashian—she just does it a little better than I do. As we use the tools at our disposal to look as attractive as possible, we both help create a world where the natural, instinctual drives of *Homo sapiens* are whipped into a perpetual state of overdrive.

This is a major reason you and I struggle to control our behaviors. Our instinctual circles are being "overfilled" by the sights, sounds, and sensations we encounter every day, while the blocks meant to regulate those behaviors are knocked aside by a flash flood of supernormal stimulus.

Of course we feel out of control. The same mechanisms that helped our ancestors thrive and stand out among species have allowed us, their descendants, to shape the world into something new and different. The outcome of our supernormal world is a growing pattern of addictive and compulsive behaviors in response to the stressful, supernormal world we've created for ourselves.

NOT EVERYONE IS equally prone to every kind of addiction. For example, hundreds of different genes can contribute to a higher predisposition toward alcoholism, but DNA alone won't drive you to the bottle. For addiction to take root, a person's environment and life experiences have to reinforce those genes. In other words, someone who grows up with loving, supportive parents who model healthy behaviors around alcohol will be less impacted by their genetic predisposition, while someone with a lower genetic predisposition may become an alcoholic if they grow up in a traumatic environment, or start binge-drinking during adolescence.

Remember, one reason why we have different genetic compositions is that life hedges her bets. I've never felt the desire to drink every day, but as we discussed above, I struggle with food in a way that most of my friends don't. This was a great system, socially speaking; if an extended famine struck, overeaters like me would survive longer thanks to greater caloric reserves in our fat deposits. Today, however, we've created a world in which calories are available with basically no physical effort. Life's strategies have been overwhelmed, and we now face an obesity epidemic, which has spawned a far-reaching diet and fitness media culture, and, in turn, driven a rise in disordered eating.

Most people don't understand what addiction is, and that makes it hard to have an informed conversation about it. I've heard hundreds of people talk about their "addiction" to their iPhone, but very few people are actually addicted to their screens—in the same way that few people who like their homes tidy fit the clinical definition of "OCD." The popular misuse of mental health terms makes it difficult to have the kinds of conversations we need to have about supporting people with real mental illness.

That means we can't simply talk about supernormal stimulus through the lens of addiction. We also have to talk about *compulsive* behavior. Compulsions are more common—almost everyone engages in compulsive behaviors today.

A compulsion is a repetitive, ritualistic behavior that a person performs without rational motivation. I'm a compulsive eater—I often seek out and consume food without thinking about it. Over time, a compulsion like this will form neural networks in our brains that bypass our rational capacity—which is why compulsive behaviors are difficult to control with "willpower."

We engage in compulsions because they offer relief from anxiety. An afternoon donut, looking at Instagram every fifteen minutes, and plucking the hair from your arms all offer an escape from a moment of anxiety or boredom. That means almost any behavior that offers some relief from anxiety can become a compulsion, whether it's food, sex, video games, or even something as benign as reading.

Of course, addictions also offer relief from anxiety. But addictions are more than just an inability to discontinue a harmful behavior. When you deny an addiction, some level of withdrawal symptom will appear. That's not true of compulsions. Addictions often (but not always) have a cycle of escalation, while compulsions can offer patterns of stable activity over years (or decades). Both, however, have the potential to become serious and can interfere with daily living.

Which brings us to porn.

THOUGH EXPERTS DEBATE the specifics, there's no doubt that an overwhelming majority of men in the developed world watch porn on the Internet—at least 70 percent, and possibly higher than 90 percent. But while porn has been a man's game, historically speaking, women and other non-male genders are gaining on their male counterparts every day. Pornhub, the website that streams the most porn, says that 29 percent of their users are women. Younger women are much more likely to look at porn— and based on trends, it won't take long for women to reach parity with men.

I hate talking about porn, because I don't want people

to assume that I'm moralizing. Religious and political conservatives talk constantly about the dangers of porn, but they do so by shaming people about their sexuality and sexual desires, projecting immorality on normal aspects of human behavior. What I see in the data has nothing to do with moralizing—I don't care what consenting adults do with their bodies. But porn is useful to talk about here, because it helps us understand what media stimuli are doing to us as a species, and why it matters.

People watch porn to masturbate. Sure, there are exceptions, but most people begin a session of porn viewing with a goal in mind: autonomous orgasm. Orgasms are powerful. When viewed in brain imaging studies, they look a lot like the high produced by heroin—which makes orgasms a powerful conditioning tool. They feel great in the moment and provide lasting emotional benefits afterward, so is it any wonder that many people who watch porn do so compulsively?

When researchers compared the brains of control groups against the brains of compulsive porn users, they found compulsive porn users had higher reward-circuit activity than their peers when both groups were exposed to porn. That enhanced reward created a craving for porn, even when the person's initial sexual desire wasn't any higher than his or her peers.

In some cases, research shows that this compulsion has serious consequences. A not-insignificant number of men experience sexual dysfunction after developing a compulsive or addictive relationship with Internet porn. And let's be clear: What is depicted in most porn isn't normal sex.

The people performing in porn are professionals. What viewers see on screen doesn't set realistic expectations for what's possible during intimacy. But that's not how most of us treat it. No one watches the Super Bowl and goes away expecting to throw like Tom Brady, but studies tell us that some number of people watch porn and then feel disappointed when their own sex life doesn't match what they see on screen.

Porn is pizza for your sexual brain. It's a stimulus much more powerful than what we find in nature, and it leads millions of people into unhealthy behavioral patterns that threaten their relationships, their jobs, and their sense of self-worth.

As I look at our relationship to porn, one thing is clear to me: Modesty and purity culture don't help. It's right in the data. Certain religious groups are most likely to promote abstinence and modesty as solutions to porn addiction: Evangelical Protestants, biblical literalists, and those with higher church attendance. According to research, those groups also search for porn more often than other groups, while the religiously unaffiliated predict lower rates of consumption.

In addition, the abstinence-only sex education put forward by religious people in the United States has been positively linked to higher rates of STIs and pregnancy among teens. Purity culture creates shame around the body and sexuality, and one consequence is that millions of women are unable to achieve orgasms—even in marriages approved by conservative religions. Purity culture says that men's sex drives are uncontrollable, and women need to cover their bodies to "help their brothers avoid sin." Is it any wonder that sexual assault and harassment are so common when we tell young boys that their sex drives are too powerful to resist, and that it's the woman's responsibility to help them avoid temptation? Or when people—religious or not—ask "What was she wearing?" after a report of sexual assault?

Shame creates hiding, not wholeness. From porn searches in zip codes with a lot of religious conservatives, to women who are ashamed to enjoy their own bodies, American culture's approach to sexuality is deeply broken. Shame creates taboos around sexual desires and drives, and taboos prevent people from reaching out to their communities for support and discussion. And this phenomenon extends *way* beyond our relationship with sex.

Soon after I started my podcast *Ask Science Mike,* questions about drugs, relational conflict, and struggles with mental health started to come in by the thousands. I couldn't understand it. Why were young people all over the world pouring their hearts out to a middle-aged man living in a small southern city? Then it hit me: They have nowhere else to go. Most communities shame people for curiosity about taboo topics—or overcompensate by encouraging behaviors that are risky or dangerous. People weren't asking me for advice because I was an expert. (I'm not.) They were asking me because they trusted I wouldn't judge them or try to coerce them into viewing the world as I do.

Flawed science and outdated ideas about addiction and compulsion have framed our lives in a way that leads us to feel trapped by our own desires, medicating our shame with behaviors that become destructive. We've trained people to view their behaviors as personal, moral failures. But that view isn't just limited—it's often flat-out wrong.

I learned about that from rats taking drugs.

You've probably heard that addiction is born out of "chemical hooks" in our brains. That theory comes from research done with rats in the 1960s, '70s, and '80s. Scientists would put rats in cages called Skinner boxes and offer them levers that yielded rewards (like a food pellet) when pressed.

When scientists started testing the effects of different drugs, they'd use Skinner boxes too. Only in this case, pressing a lever would yield a dose of a drug—say, an opiate—into the rat's body. The catch was that the rats were not only held in solitary confinement in a cage, but also tethered to the ceiling with a surgically implanted needle in their jugulars.

Gosh, if I was trapped in a room alone, strapped to the ceiling with a needle in my neck, and I could press a button that got me high, I bet I would press it too. But that's not how the results of these experiments were interpreted. When researchers saw that rats trapped alone in cages would press levers often enough to achieve high—even lethal—doses of drugs, they interpreted it as a sign that these drugs were too powerful to resist. I believe you can draw a direct line from that research to the famous TV ad where an egg sputtering in a frying pan represented "your brain on drugs." That spot and the cultural beliefs it represents also helped fuel the war-on-drugs campaign in the United States.

Almost forty years ago, a psychologist named Bruce K. Alexander looked at this research and didn't buy it. He knew that rats were social, inquisitive creatures. Locking them alone in cages offered less insight about the effects of drugs than it did about the impact of *solitary confinement* on rats' brains and behavior. To prove it, Bruce did something beautiful.

He, along with a team of researchers, built a rat park. It was an open area full of platforms, wheels, cans, and other objects rats enjoy interacting with. As with earlier experiments, rats in Rat Park were offered a choice between clean water and water laced with opiates. But when rats in the rat park tried the opiate-laced water, many of them didn't like it. Among the ones that did, their consumption rates were much lower than were those of the rats trapped in cages alone.

A beautiful picture appears here: When rats had a choice between social interaction and chemical stimulation, social interaction won. Only when rats were deprived of normal social interaction did they display signs of addiction.

Rat Park has sparked a reimagining of what addiction means. Some have said the "chemical hook" notion is baseless, and argued that it's best to view addiction exclusively through the lens of socialization. I think both camps overstate their case.

The Rat Park research had some serious methodological flaws. There's some missing data on drug consumption thanks to a broken meter. The social rats were allowed to mate, while the isolated rats didn't have the option. There have also been some researchers who tried to reproduce the findings of the original experiment but weren't able to.

In other words, I'm not saying that substances like opiates aren't powerful—they absolutely are. I'm saying that the insight from experiments like Rat Park is that you can't view the chemical action of drugs in isolation. Social factors play a powerful role in how humans deal with addiction and compulsion.

This is why we can't look at the behaviors we struggle with exclusively—or even primarily—as personal, moral failings. Drugs, porn, processed foods, all of these pull on us in the same way strings pull on the joints of a marionette. There's no question that social support and interaction are helpful for people dealing with addictive or compulsive behaviors.

If you struggle with addiction or compulsion, take some solace. There really are factors outside your control that contribute to those behaviors. You'll never be able to wrestle those demons away through willpower alone. There's no amount of prayer that will free you, and no amount of self-control that will heal you. If we want to sever the strings that pull us toward behaviors that rob our lives of health and joy, we've got to lean into what research shows us is actually effective.

When you're facing a compulsive behavioral loop, don't just struggle with your will. It's far more effective to interrupt and refocus. I've learned that staring at a slice of pizza in a display case is a battle I've lost before it begins. But if I interrupt that focus and decide to walk around the airport instead, I find I skip the slice almost every time.

Finally, punishment and prohibition are really ineffective. They train people to hide their behaviors, not to alter them. Social acceptance and support, on the other hand, are effective counterpressures to the anxiety that fuels our compulsions. This is why it's far more effective to offer education, open conversation, and professional support when people are struggling with unregulated behavioral patterns. Openness in community is a salve that helps us heal.

This isn't hypothetical. When my kids were in preschool, I was so stressed out by the demands of parenting and my career that I coped by eating fast food from a bag every day. I couldn't wait for lunch—I would leave the office for a few minutes and eat some hot food in my car, which made me feel better. You can see where this is going. Fast food became a compulsion that relieved anxiety, and I engaged in it thoughtlessly.

The outcome was predictable: I put on a hundred pounds. Less predictable was the state of my car. I was too embarrassed to walk into the office or my home with a McDonald's bag, so I'd leave it on the floorboard of my car. That cycle would repeat until there were dozens or more bags in my car, filling the floorboards. I would park my car away from everyone else, and then sneak out of the house at night to purge when the pile got unbearable.

I felt ridiculous, and tried every kind of behavioral

management tool at my disposal. Nothing worked. At least, until one day, when my friends Stratton and Colleen asked me to lunch. I was the only one with a car that could seat everyone, but I told them my car was too messy to ride in. Instead of accepting my deflection, they told me they'd help me clean it out.

I was mortified.

But I was also too passive to resist. So, here we went, out to remove an unimaginable heap of paper bags from my old, beat-up Saturn Ion. What I remember is the lack of judgment I felt from either of them. We quickly cleaned out the car, tossed the trash bags, and then went off to a fun lunch together.

Something about them seeing my car and not thinking less of me was freeing. It made it easier to toss out a bag when I got back to the office—if dozens of bags didn't result in social rejection, how could one? More than that, I found I really preferred sitting down with friends for lunch to eating alone in my car. I started to prioritize lunch with other people.

And that started me on a cycle of healthier eating and exercise that not only kept my car cleaner, but led to me running a marathon and losing a hundred pounds. Acceptance beats shame every time.

A Concrete Suitcase

The Secret Life of Emotions

Hillary screamed, and everything in my vision went dark. It felt like a circuit breaker in my brain had tripped. I couldn't see anything, and the sounds around me faded, like listening to people talk with your head underwater.

It was late summer 2018, and we were holding a retreat for thirty-two men in a house in Beverly Hills: my friends Hillary and Michael, and me. (We cohost *The Liturgists Podcast*, along with another friend, William.) We'd gotten the idea to host a retreat about masculinity in light of the #metoo movement and sky-high suicide rates for white men in America. Hillary is a licensed therapist, so we spent most of our time doing psychological exercises in a group setting. One of those activities was a psychodrama, also called a "sculpt," in which the group reenacts an event from one person's past in a safe, supportive setting.

It's not my story to tell—but I will tell you that what was shared had been a source of considerable trauma for this man. An empty chair sat in the middle of the room, a stand-in for his abuser. People around the room were telling the chair, as strange as that sounds, that what it had done was wrong, and that it made them angry.

I was on edge—the man's painful memories had echoes in my own life. But it was encouraging to watch how much the exercise benefited people in the room. You could see healing all around. Tense postures and severe faces gave way to tears, hugs, and mutual support.

But then, Hillary asked the protagonist of the sculpt, "Do you mind if I say something?" When he nodded, she turned toward the empty chair and screamed—a primal evocation of rage that I'll never forget.

That's when something inside me broke, and I couldn't see anymore. I'd like to say Hillary's scream scared me, but I didn't really feel anything at all. Instead, I hallucinated that I was in a cartoonish spaceship, traveling around our solar system. Yes, I'm completely serious, and no, it didn't make sense to me at the time. It turns out I had a particular type of panic attack, one I'd never had before.

When I came to, I didn't feel right for hours. I just wanted to run away from everyone—even my close friends. Being around people felt threatening, inherently unsafe. I wished I could leave the earth behind to wander the stars alone.

. . .

MY STORY IS an extreme case, and we'll return to it later. But I'm guessing there have been moments in your life when you, too, were surprised and overwhelmed by your feelings. As we discussed in a previous chapter, fear and anger are the most powerful and ancient emotions in our brain, and they often pop up in ways that make us feel out of control. I'm reminded of YouTube videos where someone jumps out of a trash can to startle their friend, only to get punched in the face. Fear isn't the only feeling that can overpower us. People all over the world are living through an epidemic of anxiety, for example, and younger generations experience it even more than their stressed-out elders.

Fear and anger make sense as evolutionary adaptations. In chapter 2, we learned that anxiety emerges from the interaction of brain structures that control fear and planning for the future. In chapter 3, we saw that many of the behaviors that frustrate us the most—compulsions and addictions—happen in part as a mechanism to manage anxiety. But why do our brains create anxiety in the first place? And how can we address the source of anxiousness instead of the symptoms?

When I think about feelings, I sometimes imagine a magical suitcase. We carry this suitcase everywhere we go, but not all of its contents have the same "weight." Some feelings are heavy, like when we feel sad or overwhelmed. Other emotions, like joy and optimism, are so light they almost carry us. And then there are the times when our magical suitcase does truly strange things, pulling us in directions we don't want to go. Anger can "pull" us to say things we later regret. So can jealousy. Sometimes it feels like the suitcase is full of concrete, and we can hardly drag it around.

I get thousands of messages each year from people whose feelings frustrate and confuse them. They ask me how to cope with their bitterness toward family members who have rejected them. They tell me about going to a job every day that they hate, and how frustrating it is that they can't find the will to seek out something else. Many people tell me about depression, bleak and endless. They even tell me when they feel like taking their own lives.

Our complex and confusing relationship with our emotions is a major, life-and-death concern. But I also think we approach our feelings with the wrong expectations. We've been trained by our culture to believe that happiness is the best feeling, and that life is about trying to achieve it. (Hell, the "pursuit of happiness" is one of the three inalienable rights listed in America's Declaration of Independence.) This expectation runs deep. Psychologists understand that a few "basic emotions" seem to be hardwired into our brains, and happiness is indeed one of them. It's the most popular and beloved of human feelings—really, the only one that's universally loved.

But what if happiness is overrated, and other emotions like anger, or even sadness, are misunderstood? What if our distaste for these other emotions is making us miserable?

. . .

THE WAY I relate to my feelings started to change radically at that retreat with Hillary. Later that day, she taught us a cutting-edge model for understanding emotions, which also has a super boring name: Accelerated Experiential Dynamic Psychotherapy (AEDP). The model, and its corresponding therapeutic approach, was first developed in the early 2000s by a psychotherapist named Diana Fosha, who specializes in abuse and attachment trauma. AEDP is meant for psychologists and therapists—I mean, what marketer would come up with the name Accelerated Experiential Dynamic Psychotherapy? But stick with me, the insights stemming from this model are incredible. I'd even call them life changing.

The model begins with something called *core state.* What is core state? That's the frame of mind humans inhabit when they're at ease. Core state is when you feel calm, the kind of calm that makes you curious and centered. In core state, your mind is clear, and you're able to observe and participate in the world with mental clarity. Core state is wonderful, and I think most modern people seldom experience it.

But, remember why you have a brain. It isn't so that you can experience core state, or happiness. You have a brain *so that you can survive.* Your brain's job is to map the world and help you navigate it, and your emotions contribute to this goal by serving as rapid decision-making tools. When your brain perceives something happening in the world around you, your feelings get activated in a way that motivates you to act. (We call this "arousal" when talking about the brain.) When that happens, you leave core state and go into something called *core affect.*

Core Affect
(Anger, Fear, Happiness, Sadness,
Disgust, Surprise, Sexual Arousal)

Core State
(Calm and Curious)

Humans can move from core state to core affect in response to internal or external changes, and return to core state as emotions are processed.

Our brains come "wired from the factory" with certain emotional responses. Every person in every culture is capable of these basic emotions: anger, fear, happiness, sadness, disgust, surprise, sexual arousal. Of course, depending on which researcher you talk to, the list of "basic emotions" will vary. For our purposes, the list above has widespread acceptance in the psychological community.

Long before our ancestors had the capacity for rational thinking, emotions served as a mechanism for our brains to make decisions quickly. Anger responds to an immediate threat by preparing us to fight. Fear tells us that there's probably something dangerous nearby, and we should seek safety. Happiness is the opposite. It tells you every-thing is great, and you should keep doing whatever you are doing: eating, socializing, gazing at the sky. Happiness is a reward, and we're meant to seek it. Sadness, on the other hand, tells us something is wrong, but not in the im-minent manner of fear or anger. We feel sad when some-thing or someone we value is missing, or when we experience hurt.

Disgust is vital, and certainly "hardwired." Its job is to keep us away from poisons and communicable disease. Another emotion, surprise, is a way of putting the entire nervous system on alert. Something important happened, but it happened so quickly that our nervous system doesn't know if it's good or bad yet.

Sexual arousal is different from the other emotions in that it changes over time—the sexual desires in an infant are *very* different from those of a teenager. The point here is that sexual desire isn't something we learn. It's a built-in function of the human animal, like other basic emotions. Its job, of course, is making sure humans make more hu-mans. It's nice for bonding too.

Let's try something. I'm going to list the basic emotions again. As I do, I'd like you to pause and reflect on each feeling and see how you feel about it. If the answer is "nothing," I want you to picture yourself expressing that emotion to someone else, and then someone expressing that emotion to you.

Fear

Anger

Happiness

Sadness

Disgust

Surprise

Sexual Arousal

Which of those words provoked a response from you? What did you feel? I bet most people can't read through that list without judging some feelings as good or bad—or at least better or worse than others. While all of our core emotions are healthy when processed fully, few people make it to adulthood without picking up cues that some of their feelings are wrong.

And this starts *really* early. Infants have emotions that are just as powerful as an adult's, but they lack the capacity to regulate those emotions or change their circumstances. A newborn person is a tiny emotional dynamo, and babies can be exhausting to be around because of that. But even in the first months of life, our brains start observing the facial expressions, body language, and vocal cues of those around us for signs of affirmation, withdrawal, and care.

Anger is an easy example. When a small child gets angry, the response of others will shape how the child relates to her own anger for the rest of her life. Maybe her peers rebuke her when she gets angry, or perhaps a parent withdraws. Maybe her caretaker even responds with a larger, more terrifying display of anger than the child's minor tantrum.

When that happens, we learn to inhibit our basic emotions. To use the language of the AEDP model, our core affect moves to an *inhibiting affect*. Inhibiting affects are feelings like anxiety, worry, guilt, and shame. They allow us to escape any core affect that feels "unsafe" to us. Why would our own emotions feel unsafe? Usually, it's because when we were children, our parents, our peers, or something we saw in media showed us that the emotion wasn't allowed. Here's a picture to help you visualize it.

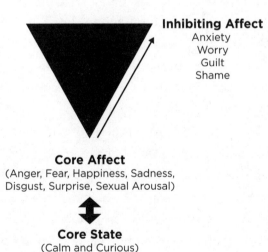

Inhibiting Affect
Anxiety
Worry
Guilt
Shame

Core Affect
(Anger, Fear, Happiness, Sadness,
Disgust, Surprise, Sexual Arousal)

Core State
(Calm and Curious)

Our emotions get bypassed by inhibiting affect when we learn from social cues that our feelings aren't acceptable.

As a kid, I got beat up all the time by my peers. When I would cry in response to other children mocking me, some of the bullies at my school would escalate to physical violence. At first, I would cry harder and louder, but it didn't take many rope burns, "swirlies," or "purple nurples" for me to learn that an open display of sadness wasn't safe at school. I felt ashamed when I would cry, and guilty for displaying weakness that allowed me to become a target.

I bet you have a story like that. Maybe not with bullying and sadness, but some other moment from your past when you learned that an expression of emotion wasn't allowed. And that's not always bad! Our response to another child playing with our favorite toy might be to punch them in the nose. In that case, the social process of restricting emotional expression is a good thing. But more often in our culture, our basic, healthy emotions are met with a social reinforcement to repress them. Instead of communicating our basic emotions, we learn to stop them by feeling anxious, worried, guilty, or ashamed.

I've been on a multiyear quest to eliminate the guilt and shame that I feel. When I notice I feel guilty, I ask if there is a way to make amends. Then, I do it. When I feel shame, I pause and ask myself where the shame is coming from, what belief is driving it. Invariably, I will find a moment from childhood when someone was unkind to me, or a place where some unexamined tenet of religious fundamentalism still lingers in my mind.

I must say, I've been quite successful in confronting those stories and releasing shame. But instead of feeling relief, I've found that I feel anxious and worried way more than I used to. As an Evangelical Christian, I've used guilt and shame as my primary tools for avoiding my feelings. Liberating myself from those inhibitions has been incredible, but since I never confronted my basic discomfort with anger or sexual arousal, anxiety has "picked up the slack" in helping me avoid the feelings I'm not comfortable with. I'd never felt anxiety before I evicted shame from my life, but now I'm making up for lost time.

Inhibiting affects help us regulate the feelings that our families, friends, and our surrounding culture reject us for. But because they feel bad, they motivate us to seek relief. When this happens, we move from inhibiting affect to what's called *defensive affect*. A defensive affect is anything we feel or do to escape anxiety, worry, guilt, or shame without processing the original emotion or stimulus. You can see some common ones in the figure on page 78.

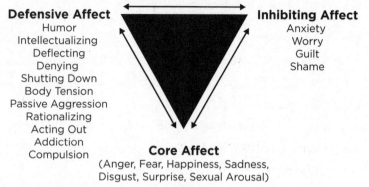

Defensive Affect
Humor
Intellectualizing
Deflecting
Denying
Shutting Down
Body Tension
Passive Aggression
Rationalizing
Acting Out
Addiction
Compulsion

Inhibiting Affect
Anxiety
Worry
Guilt
Shame

Core Affect
(Anger, Fear, Happiness, Sadness,
Disgust, Surprise, Sexual Arousal)

Core State
(Calm and Curious)

Inhibiting affects are unpleasant, so we learn to cope with defensive affect. Defensive affect helps with coping, but not with processing the original emotion and source stimulus. Often, we get "stuck" in a cycle of inhibiting and defensive affects, restricting our emotional range while also preventing us from operating in core state. Our "home base" becomes an inhibiting affect, like anxiety, and we mask that from ourselves and our social groups via defensive affect.

When I first saw this list, it hit me hard. Hillary was explaining the model to everyone at our retreat, and as she listed the defensive affects, it seemed like she was just naming the aspects of my personality I like the most. "Wow," I thought, "I'm really messed up, and I'm in the middle of writing a book about how to accept yourself and heal."

I immediately felt anxious. What if I'm not the right person to write this book? Can I afford to give the money back to my publisher? My heart galloped in my chest, and pin-prickles danced on my neck as a chill ran down my spine.

So I made a joke. I said, "Intellectualization and humor are my entire public platform." Everyone in the room laughed, and I felt a lot better.

That's the model. Something in the world (in this case, learning about the model itself) made me have an emotion. But childhood bullying taught me that fear isn't okay, so I felt anxiety instead. Anxiety feels bad, so I intellectualized the problem (which reduces emotional responses by forcing resources toward the rational structures in the brain), and then made a joke. Humor helped me transform my fear of social rejection (I'm a failed author) into a social win (I'm a funny podcaster).

That's an important point: Defensive actions aren't inherently bad. Neither are inhibiting affects. The unique way that you inhibit your emotions, and the unique ways you defend yourself from feeling bad, are not just okay. They are crucial for all of us in helping to navigate our lives and the world around us. Our defenses become a problem only when they trap us in the cycle of inhibiting and defensive affects. Sometimes we can spend so much time spinning around the triangle that we miss out on experiencing our feelings as they were meant to be felt.

I USED TO be afraid of crying. I was so afraid of it that I taught myself to never cry—and it worked.

This is a little embarrassing to tell you, but I had a

pretty elaborate set of mental constructs for controlling my feelings. Long before Pixar's *Inside Out,* I would visualize a control console in my brain that allowed me to manipulate, or even deactivate, my feelings.

It went like this. At nine years old, if a big kid had my arm bent behind my back, and he told me he wouldn't stop until I stopped crying, I would retreat into this control room and look at the dials and gauges to see how I could stop crying. I'd see the pain gauge was high, and little critical alert lights were flashing, and that the pressure from that pain was also flowing to other systems, like my breathing rate or the tension of my muscles. I'd start making changes, slowing my breathing and releasing muscle tension. If I couldn't change the source issue—physical pain—at least I could control how my body responded.

Around age twelve, I discovered the panel had a new feature: a big red button labeled "KILL ALL." Before you panic, KILL ALL wasn't a signal for me to go on a psychotic rampage. I was really into computers at the time, and you can use the system's task manager to "kill" any processes that are misbehaving and causing problems. KILL ALL was my way of doing the same with my emotions—an emergency measure that would get rid of all my feelings within a minute or so.

The problem is that I used this control so often, it became automatic. Even when I needed to cry, and there was no one around to beat me up, the most I could manage was a single sob, or a tear or two, before my body shut the feelings down. As an adult, I've been to funerals for loved ones where I can't cry. I just feel a growing pressure, kind of like indigestion, but in my emotions instead of my gut.

It took lots of therapy for me to learn to cry. By then I was in my mid-thirties, meeting with a therapist who spent most of her time comforting me and telling me I was normal. One week, when she asked me what stopped me from crying, I paused for a moment, and then told her the truth: "I'm afraid that if I start, I will never stop."

I THINK THAT'S how lots of people feel about their feelings. We're afraid that if we get sad, angry, or afraid, those feelings will overwhelm us, and we'll shatter.

But that's not how feelings work. Our feelings aren't here to break us. They're here to help us—even heal us. Our feelings can be powerful, especially when we repress them for years or decades. But when we let our feelings happen in response to events in our lives, they don't crash over us like a tsunami. No, they wash over us like the kind of warm, gentle waves I played in as a kid on Florida's Gulf Coast beaches.

Feelings are meant to have a wave action. They naturally progress, crest, and recede. But if we've been conditioned to avoid a particular emotion, we'll jump to an inhibiting affect or defensive affect before the cycle can do its work. Here's a picture to help you visualize the cycle:

A Healthy Emotional "Wave"

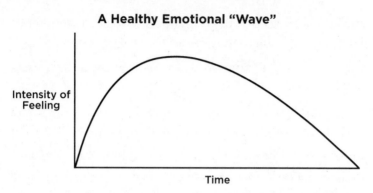

Our feelings can come on quickly, but in a supportive environment, we can process them fully. When this happens, our feelings taper off and fade to a resolution that returns us to core state soon after they feel like they are "too intense."

Emotional Bypass

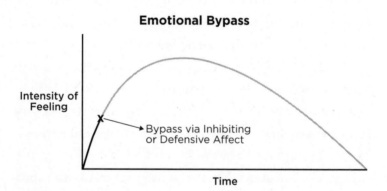

When we learn to bypass our basic emotions, we never experienced the relief that comes from processed emotions, and instead fear their intensity. That can look like cracking a joke, changing the subject, or verbally processing an experience as it happens (among others).

Many of us lack the emotional literacy to understand what we're feeling in moments of stress. We often don't know what emotion we're having, much less how intense it will be if we don't suppress it. I've found it helpful to think of feelings as body sensations. What am I feeling? And where? We'll talk about this later in the chapter, but these questions can help us understand what our feelings are trying to tell us:

> **What** are you feeling? (Pressure, warmth, tingling, burning, cold, etc.)

> **Where** are you feeling it? (On the top of my head, in my chest, in my shoulders, in my belly, etc.)

> **How big** is the feeling? (If the feeling were a circle, describe how big that circle would be.)

Answering those questions can help us move from anxiety ("Oh no . . . WHAT IS HAPPENING?") into a more direct relationship with the emotion that's causing it. One of my favorite things to do these days is to sit there and notice when I feel warmth in my eyes and pressure in my chest. I've learned to recognize those sensations as the products of sadness—but now, instead of trying to shove them down, or run from them via a defense, I just wait. Soon enough, I cry. I don't even need a commercial about kids leaving for college anymore. I can just cry and let the emotions build.

Then, just when I think I'm going to lose myself in a sobbing fit, the wave starts to pass and fade. On the other side, I'll often realize what made me sad. Many times I just miss my mom, who now lives a continent away from me.

Other times, I'll realize that a person said something that hurt me earlier that day. Either way, the feeling of sadness doesn't break me; it teaches me. I'm learning to trust it.

Of course, those of you who are paying attention may have questions here. "Hey, Mike, didn't you open this chapter with a story of your emotions overpowering you?" Yes. I did. You're right about that. But the story I opened this chapter with is not about core affect and a healthy emotional response curve. That story is about a panic attack—one brought on by trauma. For people dealing with the impacts of trauma, the work of learning to know and trust your feelings is an even more difficult endeavor.

HAVE YOU EVER heard a storyteller stop in the middle of a scene and say that time seemed to "slow down"? Of course you have. People often say this when talking about car accidents, and you also hear it in accounts of disaster situations. I've had moments like that. When I was a teenager, someone pulled a gun on me. I remember watching the weapon rise toward my face in slow motion, and how absolutely massive the barrel seemed.

Researchers tell us that we probably don't experience time slowing down in these moments. Instead, when our nervous systems believe we're in imminent danger, our brains focus their resources on recording as much sensory information as possible. These encoded memories are incredibly vivid compared to our normal memories, which are hazy and indistinct at best. So it seems like our perception is of time slowing down, when really we just use a different type of memory formation to give our reflexes and cognitive abilities the best shot at helping us survive. Moments of serious danger create high-resolution snapshots in our brains—and those snapshots stick around.

Now remember, our brains haven't been around as long as our nervous systems. In fact, your body is really smart, even without your brain. Your digestive system has something like 500 million neurons—which isn't much compared to the 86 billion neurons in your brain, but damned impressive considering there's something like 530 million neurons in a dog's brain. I'm not suggesting that your intestines are as smart as a dog, but I do think your gut (aka the enteric nervous system, or "second brain") is smarter than we give it credit for. In fact, we've found that if you sever the nerves connecting your digestive tract to your brain, the enteric nervous system will go about its business of handling digestion just fine. In those cases, it's the brain that shows the most signs of distress.

To understand how trauma shapes our every moment, we have to realize that our feelings aren't just something that happens in our brains. The same neurotransmitters that make us feel down also impact the neurons in our guts and interfere with our digestive processes—which is why digestive issues often coincide with stress or depression. It's also why people with a fear of public speaking will rush to the restroom before they can walk onstage.

The nerve that connects our guts and our brains is called the *vagus nerve*. It's a bit of a rock star in the world of your nervous system. That's because the vagus nerve connects key structures in your brain, like the medulla oblongata, to your heart, your diaphragm, and, yes, your enteric nervous system. Much like the crocodile-dog-human team in your head, the nerve systems in your body respond to stressors in different ways and at different speeds.

To understand how this works, doctors and mental health professionals use a model called the *autonomic nervous system.* In this model, your involuntary body functions are governed by two opposing systems: the *sympathetic* and *parasympathetic* nervous systems.[*] Your parasympathetic nervous system takes control when everything is going well. I call it the "digest and chill" nervous system. It tells your heart to slow down, your muscles to relax, and your pupils to constrict so your retinas aren't saturated with too much harsh light. Your parasympathetic nervous system kicks in when you have an affirming interaction with other people, enjoy a good meal, or sit down in a comfy chair.

Meanwhile, your sympathetic nervous system is type A all the way. It exists because the world is dangerous. When your brain-body system detects possible danger or a need for rapid reactions, it releases adrenaline into your bloodstream, fires up your heart, and initiates the body's fight-or-flight systems. Like Neo in *The Matrix*, your sympathetic nervous system has the power to "slow down" time by changing how your memories are encoded.

Here's why that's a big deal. When the sympathetic

* Don't get too caught up on the phrase "nervous system" here. I'm not talking about two literally separate nervous systems. You can't separate the sympathetic and parasympathetic nervous systems into two piles of brain parts. What we're describing here is an interconnected set of physical features and reaction systems distributed throughout your brain—and your body.

nervous system fires up, it can take ten to twenty minutes for the parasympathetic nervous system to calm your body back down. Let's say you thought a friend said something that really embarrassed you, but it turns out you misheard them. Many parts of your brain, especially the neocortex, can accommodate that information quickly—so quickly that you'll reach cognitive resolution before the sympathetic nervous system has finished doing its thing. This is why you'll sit there wondering why your cheeks are flushed, and why you still feel "hyped up" ten minutes later. It's confusing. We all know how this feels.

Back to your vagus nerve. In recent years, scientists have developed a newer, more nuanced model for understanding how our bodies respond to escalations in our environment. It's called *polyvagal theory*. Because the vagus nerve is not a single nerve but instead a complex signaling system that runs between our brain stems and critical organs, this model tries to explain why we experience more than just "fight or flight" in response to challenging situations. After all, life isn't as simple as "it's all good" or "we have to fight," is it? Our nervous systems have to navigate a much more complicated world, especially when it comes to social interactions.

The polyvagal theory model offers a three-tiered view of our involuntary nervous system. The sympathetic nervous system still handles the "fight or flight" response when in high arousal. But, in this view, the parasympathetic system isn't just the "chill" network. Because when the parasympathetic system is highly aroused, it can also invoke a trauma response: freeze, or faint.

In safe situations, when neither the sympathetic nor parasympathetic systems are highly activated, a third system runs the show: the *social engagement* system. The social engagement system uses the ventral branch of the vagus nerve, running from your brain stem to your middle ear, to filter out background noise and help you focus on human voices.

But the ventral branch also runs to your heart and lungs, which means your brain stem, middle ear, heart, and diaphragm can respond to the tone of someone's voice several seconds faster than the structures in your neocortex can process the content of their actual words. A sharp word can literally take your breath away. How we say something matters as much as what we say, because our automatic nervous system is listening and responding to tone long before our thinking brains get the message.

Which brings me to my Gmail inbox.

TODAY, I'M A full-time podcaster—which is almost impossible to describe to people at dinner parties. Yes, I know your nephew has a podcast he does after school. Yes, I know it's strange that I do your nephew's hobby for a living, but I do. I'm an odd sort of public figure, in that people either have never heard of me or know more about me than they know about their own friends.

My friend Michael and I got into this work because we were lonely and heartbroken and suspected other people were too. One of my first decisions was to make myself available to the people who listened to our show—after all, it was the kindness of others that helped me escape the box I was in after I left the faith of my youth behind.

After making an appearance somewhere, I will stay around and talk to everyone who wants to talk to me. Sometimes that's hundreds of people, and it takes hours. People tell me they liked my talk, or share what my work has meant to them, but they also tell me things they don't feel safe telling other people. Like how they struggle with suicidal thoughts, or how they can't find the courage to tell their parents that they're gay. I don't offer these folks advice—I'm under no illusion that I can solve their problems. What I offer people is nonjudgmental acceptance and a sympathetic ear, and I guess that's rare enough in our world that someone can make a living out of it.

But events aren't the only place I try to be accessible. When I started blogging in 2012, I had a form on my website that allowed people to send me messages—directly, not through a publicist or assistant. Thousands of people used that form every month, and I read all of their notes. I've been able to respond to only a tiny fraction the last couple of years, but it has always seemed important that I read them. Those messages helped me see why my work mattered. They made me feel like I was a part of people's lives, and not a "famous person."

Most of the messages I received were incredibly kind and supportive. Other times there were notes of constructive criticism, telling me that I misrepresented some research, got a figure wrong, or missed the mark on matters of inclusion or social justice. And then there's the hate mail, often delivered in five-thousand-word diatribes without paragraph breaks. People tell me that I'm dangerous and leading people astray, that I'm a race traitor, or a "queer lover." I even get more than a fair share of death threats—though I don't take those seriously.

For years, despite friends telling me that I shouldn't be so open to public messages, the idea of making it harder for members of the public to contact me ran against a deeply held value that I couldn't articulate. Still, over time, the experience of reading those messages added up. I started to have panic attacks when I'd use my computer. Sometimes I would read a message and then crawl under my desk to cry, or just shake.

The hate mail wasn't a problem. I could deal with that. The messages that stuck with me, and the ones that kept me up at night, were when people told me that something I said really hurt them. My first book was about recovering from religious doubt, but some folks read it and were exposed to the questions of doubters *for the first time,* and it wrecked their faith. Another time, we made an episode of our podcast about LGBTQ people and the church, and failed to include a single person of color. The notes I got from that failure broke my heart. Rightfully so. I remember feeling like I wanted to stop talking altogether.

It got so bad that if I looked at the feedback forms from our events, it would send me into a dark depression that lasted for weeks, and could even lead to suicidal thoughts. The panic attacks spread. What started as a response that happened at my computer snowballed into something that would hit me in the middle of podcast interviews—or, returning to the story that opened this chapter, in front of thirty-two men and two of my closest friends at a retreat in Beverly Hills.

A FEW HOURS after my world went dark, Michael, Hillary, and I were having dinner at a Korean BBQ restaurant. I was still feeling pretty worked up from the psychodrama, but I wanted Hillary to know what happened. When my vision returned on that rainy afternoon, I fled the room and hid from everyone. It would've been hard to miss.

Through tears, I told Hillary that when she screamed at the empty chair that day, I could see her face, and the display of anger there. I told her that her scream reminded me of another I'd witnessed long ago—a memory so vivid, and so powerful, I can't share it here with you in these pages. I have to hold at least a little for myself, dear reader.

The memory of that scream from my childhood was too much for me to bear. That's why everything went dark for me. There was no fight or flight in that moment. My polyvagal system said, "That's quite enough, Mike. It isn't safe here; it's time for you to go away." And so I did. I fled to the only safe place I could reach, the innermost enclave of my mind. That particular branch of polyvagal response is called a *dissociative event*. For folks like me who've been through a lot of trauma, our brains have learned to protect us by letting us check out when times are tough.

When I opened up about this, Hillary was gracious, and Michael was supportive. Hillary told me she was sorry that her actions upset me. I got defensive—I wanted Hillary to know that it wasn't her fault—but she wouldn't be dissuaded. She was sorry that her actions had hurt me, and she didn't care to talk about fault at all. She said it was enough for us to just deal with the feelings without working through who was responsible.

Later, as I drove her to the airport, Hillary asked me, "Mike, if I gave you the name of a therapist who specializes in trauma, would you go see them?"

My defenses immediately went up. "Hillary, I've done a lot of therapy. I think I have the tools now to work through my problems."

Ever gracious, Hillary didn't swat away my objection. Instead, she said, "I know you have. Mike, you've worked so hard, and I can see that. But sometimes we need more help and support to make it through the difficulty of working through our past trauma."

I trust Hillary a lot—she has a powerful combination of a genuine, kind heart and a professionally trained, supercomputer brain. So I said yes after some reservation.

A few days later, she sent me a name and an email address. I reached out and made an appointment with a man in Beverly Hills, of all places. But this therapist isn't like any therapist I've been to before. His name is Ron, and he's a trauma therapist who specializes in Accelerated Experiential Dynamic Psychotherapy—the triangle model I was already researching for inclusion in this book. Ron is a tall, slender man with kind eyes and a gentle demeanor. His very presence makes you feel welcome.

. . .

NORMALLY I LOVE therapy, because I'm a great storyteller. When I go to a session, I tell compelling stories and cry a lot. The therapist offers some feedback, and then I craft an Amazing Insight™ that will sound powerful later when I tell it in a book or podcast. Both I and my therapist leave the session convinced that Real Progress happened today.

I have processed a lot of problems in traditional talk therapy, and seen great impacts on my well-being. A second type of treatment, cognitive behavioral therapy, has also been instrumental in changing my life. (We'll talk about it in a couple of chapters.) But trauma therapy via AEDP isn't like that at all.

Talking therapy and CBT both play to the strengths of the storytelling brain to drive change from the "top down." The therapist begins in the outer layers of the cortical brain, before working down into the emotional brain, and ultimately into the body. But Ron doesn't let me tell stories, at least not in the way that I'm used to. I can barely get more than a couple of sentences into a story before Ron stops me—the nerve!—and asks me how I feel.

Honestly, it bugged me at first. I work hard to make the emotional subtext of a narrative obvious to the listener. Besides, when I tell stories, I care less about what I feel, or even felt at the time, and more about the emotions I'm stirring in the listener. But Ron kept interrupting me. "How did that make you *feel*, Mike?" It robbed me of my powerful defensive affects, like humor, intellectualization, deflecting, and denying.

Even worse, when Ron asked how I felt, he didn't want

me to explain via thoughts. In those first sessions, I learned that he wanted me to describe what was happening in my body, and then let it happen. Think back to those emotional response curves earlier in the chapter. Ron was training me to stop abandoning ship when I felt an emotional wave coming on.

Ron has a way of asking a question that makes it seem like your answer really matters to him. Still, I was skeptical of the whole approach. Managing my feelings with thoughts and cognition was not just second nature; it was how I'd always survived. So, after I'd been seeing Ron for a couple of months, I started the session by saying, "Ron, can we have a meta-discussion for a moment?"

"Of course, Mike. We can talk about whatever you want to talk about." Ron had a warm smile on his face.

"Some of my friends asked me about how therapy was going this week, and I told them I wasn't sure if this approach works for me. I'm a skilled storyteller, but that's not something I just do onstage. Telling stories helps me reframe and investigate challenges in my life. That process leads me to insights that offer me not just new courses of action, but emotional relief."

Ron leaned forward, just a bit, his attention focused on what I was saying.

I continued, "So when I start telling a story and you stop me and ask me to focus on what I'm feeling, I mostly just feel confused. So I just make up some sensations to tell you so I can get on with telling the story. My friends say that I should be honest with you, and tell you that, or else I'm wasting my money by coming here."

"I think your friends are right, Mike. Our work together is based on honesty and trust."

"Yeah, I know. It's just this whole approach in AEDP seems strange to me. I've done a lot of talk therapy, and I'm comfortable with it."

Without missing a beat, Ron said, "What if, in those moments where you feel confused, you tell me that?"

"I can do that."

So I asked if I could start by telling a story—without stopping to check in with my feelings. Ron smiled and said that was fine. So I told him about the problems I was having with email, social media, and event feedback forms. I told him how criticism confirms some of the most deep-seated fears I have about myself: that I am a fraud, that people wouldn't love me if they really knew me, and that eventually everyone is going to figure out who I am and I will be alone. I told him that I wanted to give up if I couldn't do the work I do without hurting anyone.

When I'd finished my story, Ron offered a look of genuine sympathy and said, "Of course you feel that way, Mike. How could you not?"

His response confused me, and I guess my face showed that, so he continued. Ron reminded me that in our very first session together, I told him about the bullying I went through as a small child, and the terrible, painful things other children said and did to me. He reminded me that I told him I didn't have a real friend at school from kindergarten to seventh grade.

And he said that my nervous system remembered that. Those events from the playground when I felt time stand still were encoded deeply in my brain—a highlight reel of traumatic moments, etched in the neurons, synapses, and impulses running down my vagal nerves. My brain and body had learned how to survive that trauma, Ron said, but I shouldn't blame my nervous system if it couldn't tell the difference between a nine-year-old boy saying my chest "looked like tits" and someone in the present day telling me via email that I'm not really a man.

In that moment, I realized that the little boy on the playground was still inside me. Not just in the form of memories, or as a metaphor for my emotions, but as a system of traumatic cues buried far beneath my conscious experiences. Deep in my brain, there was a grade-school kid without any friends—curled up, sad, and alone. I didn't know it, but he had been there for decades, watching every moment of my life pass. All the while, he was taking care of me, using the ancient wisdom imparted by my ancestors to watch out for signs that someone was going to hurt me so much again.

Today, if he sees a threat, that little boy steps in, escalating my sympathetic nervous system so I'll run away from any adult circumstances that resemble ghosts from the past. If the threat is too powerful or imminent to flee, well, that's okay. Because the polyvagal system can be used for another purpose—to hide me away behind an armor of dissociation where I can be safe, detached from events, watching them unfold from inside a comforting cocoon of catatonia.

∘ ∘ ∘

HERE IS WHERE this book gets real enough to make both you and me uncomfortable. That day in Ron's office, I realized that I practice a kind of performative vulnerability, both with my friends and with the public at large. I tap into my brain's powerful undercurrents of trauma to get the thing I need most: for people to love me through my faults.

When a situation calls for me to show emotions, I drag that wounded inner boy out from his Harry Potteresque cupboard under the stairs. I can respond to other people's tears with my own, and make people feel heard and known, because every painful beat of their broken heart can be echoed by one in mine. I've used this capacity to get people to be my friends, and I've used it to get millions of people to listen to me talk about why atheists are okay people, and the church should just let gay people be gay and move on already.

And it works. It works so well. I mean, I meet former Southern Baptists all the time who are out, proud, and married because they heard a podcast we made. Nothing makes me happier than those moments.

But I'm also learning that tapping into trauma without processing it is a dangerous and dark magic. There's nothing wrong with processing and owning your story. But instead of embracing the little boy in my brain, more often I have exploited him to make others empathize with me. I felt bad about this, so one day I confessed to Ron that the entire point of my going to therapy was to figure out how to excise this wounded child from my brain.

When I said that, Ron frowned. "No, Mike, we're not here to remove that little boy," he said. "We're here so you can learn to protect him." My eyes immediately welled up with tears as I imagined my grown-up self standing between my childhood self and the bullies who tormented me. It's an image that has stuck with me, and it brings tears to my eyes even now, as I write.

Ron told me that the way to reduce the panic attacks wasn't by getting rid of the memories, but by honoring them by accepting them as part of me. A vital part of me that has not only kept me safe, but transformed me into a kind and empathetic person who doesn't turn away from the suffering of other people.

The following week, I took a radical step to protect that little boy. You can't email me anymore—there is no public form on my website. You can't tweet me, or Facebook me, or even Instagram me. There's grief in that for me, but I had to let it go in order to honor the wound in my heart that is also the gift I offer you.

It's helping. As I've retreated in small ways, and continued processing all my stored traumas, I no longer have as many panic attacks. I've even learned strategies that help me cope with the attacks as they happen. When I start to feel my sympathetic nervous system take off, I plant both feet on the floor, grounding me in this moment—not in whatever memory my nervous system is running away with. I'll take my hands and run them along my jeans, paying special attention to the feeling of the fabric. Then, I look around for an object I can fix my eyes on and study—again, drawing my attention to the here and now. If all those things don't work, or my feelings of fear continue to escalate, I pull out the big gun: cross-body tapping.

Which means I just cross my arms and tap my right hand on my left shoulder and my left hand on my right shoulder in an alternating beat. Doing so stimulates activity in both hemispheres of my brain, because I'm working at regular, frequent intervals across the midline of my body. I was skeptical at first, but it really does help me calm down. Soon, I can process my feelings instead of turning them off altogether.

Studying AEDP and going to a trauma therapist haven't just helped with my trauma. I've also learned to be less afraid of my everyday feelings (though I'm still working on that). I've learned to sit and be sad when I need to be sad, instead of deflecting or making a joke. I've also learned that when I feel anxious or guilty, those feelings are inhibiting something else. I take the time to ask, "What do I feel anxious about?" Recently, that's been the fact that this book is overdue to my publisher. I am afraid that my editor, Derek, won't like me and that my agent, Christopher, won't want to work with me anymore. I'm afraid that my first book was a fluke, and that I'm not a good writer. And all of those fears are piled on top of a deeper one: a fear that I am not worthy of love, just because all those kids were mean to me when I was a child.

Anxiety doesn't tell us very much. It's like the "check engine" light in your car, which tells you something's wrong without telling you what it is. To escape anxiety, we can't just focus on external factors. We have to look under the hood.

. . .

THERE AREN'T ANY simple answers in this chapter. If this were a self-help book, I would end it by distilling all of this complex science into some simple insights and actionable tasks. Instead, I've offered you an academic model in psychology, a bunch of biology, and some embarrassing stories that made me cry a lot while writing.

Here's the thing: There are no easy answers in the arena of human emotions. Sometimes, protecting yourself means backing off, like when I took that form off my website. But other times, growth and healing come from leaning in and pushing through, like when I told Ron I wasn't being honest with him. I don't have an easy rubric to help you decide which is which for you right now.

What I really want you to know is that I'm a successful, popular person who is also a total mess. And I tell you that because I want you to learn to love yourself—all of you.

I believe that when you learn to appreciate all your feelings, to accept them as a vital part of you, you'll begin a journey that helps you escape the cycle of spinning around the triangle of core, inhibiting, and defensive affects, and instead acquaint yourself with the clarity and calm that come from riding an emotional wave all the way through.

But I also want you to love the way you crack a joke when you're uncomfortable. I want you to have the tools of anxiety and guilt when you need them. I just don't want you to remain trapped there, lost and confused in the maze that your body built to help you survive.

The Machines Are Winning

*How Computers Domesticated Humanity—
and How We Can Fight Back*

What do the following movies have in common: *The Matrix, Terminator, Avengers: Age of Ultron, Ex Machina,* and *I, Robot*? It's easy. They're all about machines trying to take over the world. Many folks have written about our culture's fascination with apocalyptic stories. We love watching zombies, meteors, and aliens wipe out entire cities—and that assertion has billions of box office dollars behind it.

Regardless of how many tickets have been sold, I don't think many people are afraid that zombies or aliens are going to take over the planet. Large meteors have struck the earth in the past, and undoubtedly will in the future, but their visits can be hundreds of millions of years apart. Those films serve some other role, beyond the scope of this book, in our collective psyche.

But the rise of machines? People seem to have genuine concern over that. Computers are advancing rapidly in their ability to think, and it seems inevitable that at some point, their impossible speed will overtake our slower, organic computing brains. I think that's why so many summer blockbusters are about a machine uprising. It seems likely that computers will one day gain the ability to think, and when they do, humanity's time at the top of the food chain will come to an end.

I have bad news. The machines have already won, and they took over human civilization without ever becoming sentient, much less firing a shot or overthrowing world governments. In fact, we built them, take care of them, and willingly offer our life experiences to them.

Which reminds me: Alexa, how's the weather today?

ALMOST EVERYONE KNOWS this, but in case you missed it (or are too young to remember), a computer called Deep Blue beat a world-champion chess player named Gary Kasparov at chess in 1997. Then, fourteen years later, a computer called Watson beat the two best Jeopardy players in the world. Unlike with chess, playing Jeopardy required the computer to parse human speech and respond to the host's "clue" with a question. The progression from winning at chess to winning at a more complicated game like Jeopardy amazed the public and galvanized the computer science community.

Deep Blue and Watson were similar in many ways. Both machines were carefully tuned by teams of programmers, and their victories showed how human intelligence could be amplified when working alongside a computer's brute-force computation. But to give computers the ability to compete with our intuition, a new approach was required.

In 2015, a program called AlphaGo beat the top human players of a game called Go. If you're not familiar with it, Go is a game from East Asia played on a small nineteen-by-nineteen grid. Based on the rules of the game, there are more possible configurations of the game board than there are *atoms in the observable universe.*

Go is a game that is won with intuition and experience, which means computers struggle with it. When programmers tried playing it with the approaches behind Deep Blue and Watson, their computers couldn't compete with even the most mediocre of human players. Some experts questioned if computers could ever be competitive in games like Go at all.

But DeepMind, the company behind AlphaGo, didn't teach their software how to win at Go. Instead they created AlphaGo with the ability to learn how to win *on its own.* They taught it the rules of the game, and then showed it games that others played in order to learn which techniques won games. It was an impressive win for machines (and their programmers). But AlphaGo's victory was short-lived.

A new version of the program, AlphaGo Zero, dominated AlphaGo less than two years later. AlphaGo Zero didn't learn Go by watching others play. It learned by playing itself. Using that strategy, it became a master Go player in three days, playing almost five million games against itself. AlphaGo Zero dominated the game in a way that no player—human or computer—has ever done. How dominant? In its first competition, it beat the previous champion in one hundred out of one hundred games.

Machine learning is more powerful than it's ever been. But my concern in this chapter isn't how machines like AlphaGo Zero are taking the board-game industry by storm. It's the increasing—and increasingly dangerous—role these programs play in shaping the way you see the world.

Before I get into how machine learning impacts your life, let's take a moment to understand how machines learn at all. There are a lot of different strategies, and only nerds like me enjoy reading about them. For the sake of simplicity, I'll use a somewhat dated, but still relevant, model.

Imagine you want to teach a computer to identify pictures of pizza. It would be nearly impossible for you to handwrite an exhaustive set of rules for the computer to follow. We don't really know how humans identify something like pizza. From a young age, we just do it—and do it very successfully.

When programmers create a computer that can learn, they'll create a series of programs, which you may want to visualize as robots. First, they'll create a Maker Bot, a robot that creates other robots. The little robots it makes we'll call Student Bots; they're made from common parts, but assembled randomly. The programmers then make another robot—let's call it a Teacher Bot—to train all the Student Bots how to tell if a picture is pizza or not by showing them images of pizza and telling them what pictures are pizza and what pictures aren't.

After the students have trained, a new bot enters the picture. This is the Tester Bot. The Tester Bot asks all the Student Bots to take a test, using pictures they haven't seen before. The Student Bots who score an A get promoted into the next class, where they become templates for the Maker Bot's next round of students. The Student Bots who scored a B, C, D, or F are brutally murdered, and thrown into an incinerator.

Machine learning takes the selection pressures of biological evolution and accelerates them. Facebook, Instagram, Twitter, Google, YouTube, Netflix, Spotify, and Amazon are all powered by massive—and I do mean massive—computer networks running ruthless machine-learning systems. The algorithms that survive are the ones that deliver a business goal for their human overlords. That goal is usually either holding your attention (social media) or getting you to buy something (e-commerce).

Learning machines are completely amoral. They don't have a moral compass or sense of ethics. All they do is continuously and relentlessly optimize themselves toward the task they've been assigned to. For the machines behind Facebook, Instagram, and Twitter, that goal is holding your attention so you will look at ads. Those machines have followed our every move for countless, untold hours. They have learned that we like to look at smiling faces and sexy bodies.

When the machines learned we liked to look at happy,

attractive people, the seeds for mass loneliness and depression were sewn. The machines watch what we like, and then offer a curated form of reality that is not just inaccurate, but hurtful. When people look at social media, they see all their friends hanging out, smiling and having fun without them. But the reality is their friends don't go out any more than they do—and when they look at Instagram, they experience the same sense of isolation. Sociologists even coined a name for this collective sense of loneliness: the Instagram Effect.

When AlphaGo Zero learned to play Go, it discovered, and then discarded, most of the strategies employed by human players. When the machines behind social media watched us, they held on to their findings and learned that there's something we pay even more attention to than beautiful people. That something is moral outrage.

Social media has showed us a side of our friends and family that many of us had never seen. As social media moved away from a chronological presentation of information to an algorithmic presentation, the content that caught attention was elevated to the top. One of the surefire ways for a piece of information to hold our attention—and drive us to act by sharing or commenting—is to make us feel moral outrage.

Outrage has a purpose. Animals sharing space need a mechanism for knowing when to punish the selfish acts of other animals. In social animals like humans, that mechanism gets filtered through cultural norms and taught to us through participation in social groups.

Moral outrage creates activity deep in our brains. It's potent because it has a social component, and politics and religion play a key role in people's sense of identity and belonging. Outrage makes us feel like a good person ("I care! I won't stand for this!"), and at the same time it makes us feel like a part of a group. It also offers us a foil in the form of someone who is clearly wrong from our perspective.

Moral outrage isn't bad. Not at all. Human civilization wouldn't be viable without it. Moral outrage told us that what the Nazis were doing was wrong and moved the world to action. The triumphs of the civil rights movement in the United States were likewise powered in part by moral outrage. But social media offers us a supernormal form of moral outrage, unnatural in its frequency and intensity.

When someone calls out racism or sexism on social media, or offers a necessary and thoughtful critique of someone they mostly agree with, it has a strong likelihood of motivating people to act. That post can get several likes, comments, or shares early in its existence. The machines take notice, and reward the post by placing it in more and more people's timelines. Before long, a confrontation between two people has snowballed into a newsworthy event.

At that point, many people will agree with the criticism, but others' social brains will intuit that the punishment is too harsh, and perceive the people calling out the transgression as bullies. In fact, one study found that when a social media post with moral indignation goes "viral," many onlookers will perceive it as bullying—regardless of the validity of the outrage. Which is ironic, considering that the person being called out is often a powerful media figure, and the people responding with outrage are often college students, poor people, or people of marginalized identities who are directly impacted by the rhetoric in question.

We're training the machines, and the machines are training us. Social media spotlights issues of human justice that deserve our attention, but it also drives us into increasingly polarized camps by presenting us with the most extreme views of a given group. We learn to associate people on "the other side" with the most extreme voices in media—and sadly, this impression is becoming more rooted in fact. The breadcrumb trail of recommendation engines is driving people to more and more radical worldviews in a quest to hold their attention. Just look at the emergence of popular movements like flat-earthers and anti-vaxxers.

Less nefarious to society, but perhaps worse for your health, is the way the machine-learning systems at companies like Netflix or Amazon suggest another program just as you finish the one you're watching. There are no commercial breaks, and no transitions to a different show that doesn't interest you. That means millions of people "just one more show" themselves into chronic sleep deprivation.

What makes me saddest is when I consider who "wins" from this activity. It doesn't matter which political parties are in control, or which social movements gain popular attention. The big winners in machine-learning-driven social media are the companies in Silicon Valley. #Blacklives matter, #metoo, and #MAGA all make Twitter, Facebook, and Google wealthy and powerful. As do the flat-earth movement and anti-vaccination conspiracies. These companies so effectively monetize our attention—and our collective misery—that there's very little incentive for them to behave in ways that would benefit mental health and societal cooperation.

Most tragically, to me, is that so many marginalized groups put serious time and effort into social media advocacy thanks to its effectiveness in helping them organize together and capture national media narratives. But all their work generates impressions, which attract paid advertising that ultimately makes a handful of billionaire tech bros in Silicon Valley even more rich and powerful, while the marginalized people doing the work get nothing. It's an economic con job of mythical proportions.

To claw back our freedom from Silicon Valley behemoths, we need to study four trends that have allowed digital media to hijack our instincts. With that understanding, we gain the ability to create strategies that let us live healthier lives in today's digital world.

Supernormal Screens

In the last one hundred years, humans have gone from primarily looking at the world around them to primarily looking at rectangles of different sizes, presenting us with information. The depth and breadth of information available almost anywhere, completely wirelessly, and in high-definition video and sound, is a profound achievement.

But, as we discussed in chapter 3, all these displays are a form of supernormal stimulus that overwhelms the survival instincts that we're born with. Somewhere along the way, engineers' quest to make displays more and more lifelike crossed over into creating colors so bright and vivid that they rarely exist in nature. I've got a 4K High Dynamic Range TV. When I watch movies in 4K HDR, I'm struck by how they seem even more real than reality somehow. It's unnerving, to the point that I've turned off a lot of the features on my TV that create that hyperreal presentation.

The forces of market economies demand this. Each phone or TV needs to be brighter and more vivid than the one you already have. Otherwise, there's no incentive for you to buy it. But this race to the supernormal doesn't end when you've bought a TV, computer, tablet, or smartphone. The competition continues as apps try to stand out among the competition, and videos try to seize your attention with the most powerful previews they can offer.

What I want you to see is that there are profound forces driving companies to create ever more supernormal displays, which in turn trains consumers to expect it. The accessibility and portability of smartphones and tablets is especially problematic. Doctors are treating more and younger patients for neck and back problems, stemming from the near constant hunched-over posture so many of us spend our time in. These superbright screens offer a welcome escape from boring or awkward situations, and we've taken that escape so often that it's hurting our bodies.

We even take these devices into our beds. The rise of smartphones and streaming video has been met by an accelerating trend toward mass sleep deprivation. People are sleeping less, and when they do sleep, that sleep is of lower quality.

Our brains have a few systems for regulating our wake/ sleep cycle. The primary system is based on light. Is it any wonder that staring into a bright light into the late evening and night makes it hard for your brain to move into a sleep cycle? For hundreds of millions of years, the sun was the sole source of bright light on this planet, and our bodies evolved to use that light as a clock. Today, we've hijacked those systems, and we're unhealthy for it.

If you take anything from this book, let it be that sleep is vital to your physical, emotional, and mental health. It's how bodies handle critical maintenance on our brains and other systems. When we talk about health, we tend to focus on diet and exercise. But eating healthy foods and getting adequate exercise won't help you much if you are sleep deprived. In addition, studies have shown that not getting enough sleep increases your risk of cardiovascular disease by 45 percent. And people who sleep more are more creative, happier, and more able to cope with stress.

If there's one way to unlock your full potential as a human being, it's getting seven or more hours of sleep every night. But that's hard to do when the machines are in your pocket and on your nightstand.

Compulsive Craving

In chapter 3, we learned that the potential reward offered by the *buzz buzz* of your silenced smartphone is a potent tool for creating compulsive behavior. The fact that the same little sound could represent social validation *or* bad news makes your brain release dopamine, and in doing so creates a cycle of craving. You can't help it; you want to know what's next. Our smartphone notifications are an irresistible siren's call for our attention, to the point that people often feel uneasy when they don't have access to their phone.

I was there when Steve Jobs introduced the iPhone, and I've spoken with engineers who worked on that device. It's hard to remember now, but when the iPhone launched, there was no App Store, and the only apps were the ones that came preloaded on your phone. Those early iPhones weren't as addicting, largely because notifications were so limited. Smartphone notifications were originally designed to let you know when you had a meeting or needed to wake up. But text messages and emails were the only kinds of communications that could notify you, and most people got a lot fewer messages back then.

But software companies pushed to get access to those notifications, and smartphones went from alluring to addicting. Make no mistake, it's normal for game developers and social media sites to keep behavioral psychologists and neuroscientists on their payrolls. These companies have learned the hidden forces that shape your behavior, but they've done so to make money, not to improve your life.

This triple-pronged threat: machine learning, supernormal screens, and compulsive craving are the battle strategies machines have used to take over our lives. We've lost the war already, and the toll on our mental health is staggering.

The Anxiety Epidemic

Though the twentieth century was an era of rapid, even accelerating, social change, some behavioral trends remained stable. Let's look at dating and sexual activity, for example. The boomers and Gen X* (terms that have no basis in science but make reasonable shorthand in conversation) have different worldviews, but as high school seniors, 85 percent of people in both generations went on one or more dates. For all the societal shifts that happened across that forty-year period, dating looked much the same.

---

* For more information on why I can't use generational labels like "boomer" or "millennial" without rolling my eyes, see Appendix B.

That's not true of millennials. In 2015, only 56 percent of high school seniors went on one or more dates—a sudden and dramatic departure from what was a stable, multigenerational norm. That shift has driven other changes: High school sexual activity is down 40 percent and the teen birth rate is at an all-time low. There's no question in my mind that the decline in the teen birth rate is a positive development, and I'd imagine the stats on sexual activity and dating would be viewed as positive by many Americans. I'd argue that this news is more ominous than it appears at first glance. To see why, we need the context of additional data points.

Millennials and the iGeneration both wait longer and in some cases much longer, than previous Americans to get driver's licenses. The rite of passage that represented freedom and independence to previous American generations seems not to interest modern young people.

I think it's related to this figure: Time spent with friends and family has dropped 40 percent since the year 2000 for millennials and the iGeneration. Of course, adolescents tend to spend less time with their parents as they grow older. But friends? Young people are spending less time among their peers than prior generations, and it's taking a toll on their mental health.

Young people today are the most depressed and anxious youth we have on record. Granted, depression has been rising among Americans ever since the Silent Generation, but among young people, anxiety has become an even more pressing problem. Among children, there's been a 20 percent increase in anxiety diagnoses from 2007 to 2012. With teenagers, 38 percent of girls and 26 percent of boys have an anxiety disorder.

The data are clear: We are in the midst of a mental health crisis, with younger people experiencing distress at historically unprecedented levels. The causes of this crisis are numerous and complex, but one factor is uniquely implicated: digital, social media in your pocket.

I've noticed people get defensive when I talk about this, so let's start with the obvious. I'm an unlikely candidate to critique our increasingly digital lifestyle, for four reasons:

1. I have significant challenges with learning and information processing that are largely mitigated when I have access to a computing device of some kind. If I had been born a few years earlier, there wouldn't have been computers in my schools, and it's unlikely that I ever would have been admitted into "normal" classes.

2. I spent most of my adult life working in information technology, starting in the help desk and ending up in the C-suite. My work revolved around helping people and organizations learn to use computing systems in a way that improved their productivity and quality of life.

3. After starting out in IT, I moved on to digital advertising, with an emphasis on social media. I've convinced companies of all sizes that they could better connect with their customers if they sought them out on social media.

4. As a podcaster, my entire business, and connection with my audience, happen via the Internet and social media.

In other words, I'm more dependent than most people on digital technologies and the Internet, and I know it. I'm no Luddite, screaming about the dangers of newfangled media like newspapers or telephones. Quite the contrary: I believe that digital technology has a profound potential to improve human life. But that doesn't change the fact that experts and researchers are finding that the devices we rely on are also having a growing negative impact on our quality of life.

HERE'S A QUOTE from Jean M. Twenge, a professor of psychology at San Diego State University, writing for *The Atlantic*:

> All screen activities are linked to less happiness, and all nonscreen activities are linked to more happiness. Eighth-graders who spend 10 or more hours a week on social media are 56 percent more likely to say they're unhappy than those who devote less time to social media.

Based on my experiences talking about this phenomenon onstage, I know many people see the above statement as an overgeneralization of the research, while others become downright defensive when they hear it. You'd be right to do so. I mean, without social media and smartphones, there would be no #blacklivesmatter or #metoo. The immediacy of social media has enabled advocates to press topics into public view that are more often swept under the rug by politicians and traditional media. Thinking about the impact those movements have had on our culture brings tears to my eyes.

The Internet can act as a great equalizer for disabled people too. I have nonverbal autistic friends who can't speak but can tweet. People who experience severe social anxieties are often better able to communicate with others via text messages or posts than they could in person.

A particularly poignant example for me is a friend I made in a virtual reality service called AltspaceVR. He told me that he is a trans man and uses a wheelchair for mobility. He liked to interact with people in VR because no one questioned him about his gender, and people didn't judge or pity him for being in a wheelchair.

I say all that so you'll know I understand the complexity and nuance surrounding these technologies. But I stand by the core notion that right now social media and smartphones hold significant responsibility for the mental health crisis of today. Researchers have found that heavy social media use increases depression risk by 27 percent, and heavy smartphone and tablet users are 35 percent more likely to have suicidal ideation. Far from providing late-night monologue material, our relationship to our devices has life-and-death stakes.

I think we've entered an era of unintended consequences. Technologies designed to liberate the human mind from mundane tasks and enable us to communicate on a personal, global scale have instead been used to disturb our sleep, destroy our productivity, polarize our politics, and drive us into compulsive behavioral patterns that steal our capacity to engage with others socially.

And it does all that while making us *feel* more productive, but actually lowering the quality of our work.

The Myth of Multitasking

When I was a child, they had only just figured out how to make a computer small enough to fit on your desk. Computers were comically limited compared to anything we use today. Those early PCs could run only one application at a time. That wasn't a big deal if you were writing a letter, or drawing a picture. But whenever you tried to work on something that required information from two or more applications, you were in for a painful process that included swapping disks in and out of the computer constantly.

Switching applications took so much time that you'd often forget what you were supposed to do in the other program by the time it opened. Thankfully, as computers became more powerful, their operating systems were updated to allow you to run more than one program at once, and switch between them quickly. This made it easy to, say, make a chart using a spreadsheet, and then drop it into a presentation in a matter of seconds. Computers learned to multitask in order to help us focus on a task that required more than one application to complete. But that capability introduced unexpected changes in *our* behavior.

When I entered the workforce, the normal way to communicate was interoffice mail. We'd actually write memos, drop them in envelopes, and get "mail" from other people in our office. But once computers could multitask, we could have our email program open while writing a letter, searching a database, AND copying files onto a server. It felt like you could get so much done. That feeling, of course, is an illusion. Modern, stressed-out workers know this in their bones, and research supports the validity of their experience.

In one experiment, participants were asked to switch between complicated tasks like sorting shapes and solving math problems. People lost time when they had to switch between tasks, and the more complicated the task, the more time they lost.

In another experiment, this time involving bilingual people, participants were asked to name two digits off a display. The catch was, the digits were color-coded, and depending on the color, the participant was supposed to name the digits in either their primary or secondary language. Of course, it took people longer to name digits in their second language than in their primary, but it also took people longer to name digits in their native tongue after having read them in their second language.

Even when people are asked to switch between tasks at a predictable interval, they perform tasks more slowly than when they stick to the same type of task. We're not natural multitaskers. There's a quarter-size patch of tissue behind your forehead called the prefrontal cortex. The prefrontal cortex is part of the brain's neocortex, and is sometimes referred to as "the CEO of the brain." It seems like the CEO gets overwhelmed when faced with the need to rapidly switch back and forth across types of tasks.

Researchers theorize that our brain's executive function has two distinct stages. Stage one is *goal shifting*, the process by which you decide to do something other than what you are doing right now. The second stage is *rule activation*, where your brain accesses what it understands about the task. Both stages happen quickly and automatically—in just a few tenths of a second—but if you switch tasks often, those fractions of a second add up. Plus, the act of switching can create mental blocks as your overworked brain struggles to keep the right set of information in your working memory. Those blocks can eat up as much as 40 percent of the time you spend working.

This is why it's so dangerous to text while driving. It's not just that you take your eyes off the road—which is dangerous enough. It's also that the time you lose by switching between the mental states of driving and texting keeps your brain from being able to respond to the environment. Those tenths of a second are plenty of time to collide with another car at road speeds.

When we incorporate messaging systems (email, texts, chat, etc.) into our digital workflows, you can see how those fractions of a second add up. If your email program or WhatsApp is open in the background, they can demand your attention at any moment. Your brain knows this and has trouble settling into the task at hand as a result.

Multitasking leaves us feeling stressed out and overwhelmed. More than that, our constantly divided attention impairs our capacity to connect with others—even glancing at your phone during a conversation distracts you and sends a message to your conversational partner that what they say isn't that important to you.

In a war to control human thought and behavior, the machines are winning. The cost is both in quality of life and in life and death. We're more lonely, anxious, and outraged than ever, and we're losing the people we love to death by suicide, on one hand, and distraction-driven road fatalities, on the other.

In George Orwell's *1984*, the government forced citizens to accept a telescreen for surveillance in their home. In reality, we ordered telescreens with Amazon Prime shipping, or camped out overnight to buy them at Apple Stores. We're in a cage of our own making.

But we don't have to give up. We can fight back—and we can win.

Fighting Back

I don't think social media, smartphones, streaming video, or even machine learning are inherently bad. In fact, I think it's possible to use these technologies in a way that helps us use computers as the "bicycle for the mind" they were meant to be, without wrecking our health.

The only way to do that is to set up boundaries with the devices and companies that are incentivized to monetize our misery. I've been on the forefront of misusing these tools, which means I've had to learn to rein them back in out of necessity. Here's my battle plan to regain our humanity.

Focus. Structure your time around working on one thing at a time, and close down any applications unrelated to that task. If you're writing a paper, just write that paper. Close down your web browser, messaging apps, and email. I find it helpful to use a timer. I work for twenty-five minutes, take a break for five, and then get back to work for another twenty-five minutes.

Of course, communicating with others probably plays a significant role in your life and work. The point is not to avoid communication, but to focus on it. I check my email two or three times a day, but when I do, I tackle everything there. The same is true for text messages—I can take hours to respond to a text, but when I do, I am completely present in my reply.

Slow down on social media. I've tried to leave the big corporate social media sites multiple times. Every time I do, I feel happier, more grounded, and more connected with others. I also feel broke, because my business suffers immediately!

Since I can't leave, I set rigid boundaries. I don't install most social media apps on my phone, so I check them like I would my email. I sit at my desk and focus, usually for ten minutes, a couple of times per day.

Whenever possible, I use a chronological feed on social

media because it eliminates the supernormal feedback loop of the happiest faces and the most morally outrageous posts. I also use open source social media platforms (like Mastodon or Discourse) as much as possible. Because they don't use machine learning to curate their presentation of content, I've found them to be an oasis of calm in the frantic world of social media.

Reduce or eliminate notifications. The biggest driver of compulsive behavior in today's digital devices is notifications. On my phone, I don't allow applications to send me notifications, other than phone calls, Delta, my bank, and GPS applications. On an average day, my phone will *buzz buzz* three to five times. I do the same on my iPad and computers. Nothing is allowed to demand my attention. Instead, I direct my attention to one task at a time.

Honestly, it's hard to describe how liberating this is. At first, I felt panicky, wondering what I was missing. News is my main addiction, and I was worried what developments were happening in the political world during the hours when I wasn't looking at Twitter. But the machines in Silicon Valley want us constantly engaged with world events, not to make us better citizens, but to make them more money. And it turns out the world keeps on spinning if I read the news only once per day. I've found I feel more able to respond to challenging situations in the world.

Giving my body an escape from the cycle of stress-inducing stimulation gives me the energy to meaningfully act through direct advocacy on the causes I care about, instead of feeling perpetually worn out and hopeless. By taking care of myself and putting boundaries on notifications, I'm actually more able to be an effective advocate for change.

Put your devices to bed. It's not enough to simply get rid of notifications. If your devices are nearby, they offer immediate relief from boredom or anxiety. To combat that, I try to follow a "one device" rule. If I am on my computer, or watching TV, then I keep my iPhone on my nightstand and my iPad in my bag. Multiple devices put us in a constant state of context switching, and I avoid that by using only one device at a time.

When I am healthiest, I also have the discipline to maintain "device free" hours. For me, that usually means not looking at any screen until after breakfast, and then putting my phone on its cradle in the evening. I like to watch TV with my family, but watching TV for too long after the sun sets is bad for sleep. So my family is learning how to have evenings together without screens. We sit outside a lot, play board games, and talk about our dreams and fears.

Of course, my experience isn't universal—other tactics may work better for you. The point I want to make is that the interaction between your instincts, the supernormal stimulation of modern devices, and machine learning is one of the biggest reasons you are so often a pain in your ass. Your life will improve the more you push back against these systems, allowing you to sleep more soundly, improve social interactions, and even create more meaningful work.

The world our species has created for ourselves is full of supernormal stimuli, from massive movie screens to pizza that gets delivered automatically with a single tap on your phone. But these external forces aren't the only factors that pester us and demand our attention in unhealthy ways.

Sometimes, the enemy lies within us. That's what I want to explore in the next chapter.

Sticks and Stones

The Destructive Power of Language

By now, you know that I was an incredibly sensitive child who endured a lot of bullying in school. When I would break down and hide from my peers in those moments, adults would often find me and try offering comfort with these words:

Sticks and stones may break your bones, but words can never hurt you.

I *hated* that phrase. I'd take a skinned-up leg over hurtful words any day. Physical pain had the advantage of offering a predictable path toward healing. What bleeds right now will scab tomorrow, and turn into a scar over time. But the pain that came from words? Those invisible scars can last a lifetime and show up in the most unpredictable moments.

To survive, I learned to bury all that pain inside, and to disconnect my thoughts and feelings from my body awareness. All that latent emotional energy is usually invisible to me, at least until something triggers it, like you saw in chapter 4. Or like the first time I heard that a mass shooting had happened on a school campus.

In 1999, I was just a few years out of high school when the Columbine High School shooting happened. This was a time when mass shootings were still rare. Everyone in my office walked around in stunned silence or disbelief. They couldn't imagine how this had happened. How could anyone walk onto a school campus and shoot students?

I knew exactly why. While everyone else was asking, "How did this happen?" I was thinking, "How did this take so long?"

There is a pain that every person who survived childhood bullying knows and understands. Every insult, every tease, and every act of abuse is another brick that grows over time into a monument of rage. The anger stays with us every moment, and we know it's dangerous. By the time I was in middle school, I took great delight in building explosives, smashing old buildings at night, and engaging in other acts of mindless destruction that let me channel my anger. I want to be clear that I never imagined walking onto campus to kill other people. But in high school, I remember imagining what it would be like to walk onto campus with a gun and execute myself in front of the children who'd tormented me the most.

These fantasies, by the way, continued long after I learned to make friends. While elementary and middle school were hell for me, I actually had a lot of fun in high school. My body went from chubby to lanky; I learned to play bass guitar, and had a pretty popular rock band. But the pain I felt inside never let me believe that these friends were here to stay—and it certainly never forgave those who hurt me the most.

That's why I had my own, private version of the singsongy refrain adults so often used to try to comfort me.

Sticks and stones may break your bones, but words can fucking kill you.

As far as we can tell, ancient cultures didn't have a word for the color blue. The Egyptians were the first people whose language included it, likely because they could produce blue dyes. Elsewhere, though, there are ancient texts that speak of the sea as being the color of *wine*, which is a really different perception from what I had the last time I was at the beach. This led a psychologist named Jules Davidoff to wonder if ancient people could see the color blue at all.

Davidoff wondered if the sensation of blue didn't exist before the word itself was invented, and he tested his hypothesis with the Himba tribe from Namibia. The Himba don't have a word for blue at all, though they have many more words for the color green than do English-speaking cultures. Davidoff composed a simple experiment: He showed members of the tribe a circle, made of squares. All the squares were green, save for one, which was a bright blue. Then, Davidoff asked the subject to identify which square was a different color.

The result? Members of the Himba tribe struggled to find the blue square. Without a word for blue, their brain's visual system never learned to differentiate blue wavelengths of light from green wavelengths. They are unable to discern blue from green, even though their eyes have the same photoreceptors that you and I have in ours.

But remember how the Himba have many more words for green than we do? When Davidoff ran the experiment again using green squares, and one with a *slightly* different shade of green, the Himba had no trouble spotting the subtle variation in green, while English-speaking people had just as much trouble finding the "different" green as the Himba had finding the blue square. I've tried this experiment myself—that damned different shade of green is really hard to find.

Think back to the opening of the book, where we saw how the terms included in a word scramble influenced how fast a subject walked when he or she left the room. There are countless experimental results like this, showing how much words impact our subconscious experiences, feelings, and behaviors.

I'm reminded of an experiment that asked people to write down the last few digits of their social security number before participating in an auction. People whose numbers had higher digits actually bid more money on items than people whose digits were lower. Advertisers know this and spend incredible resources priming people to accept the price of a given product as not only reasonable but nonnegotiable.

Words shape how we see the world, how we behave, and how we assign value. Is it any wonder that words can also lead us to question our own value? That words can even lead us to die by suicide?

WHEN ANIMALS ENGAGE in patterns of behavior that result in their death, they're generally guided by an altruistic motive to protect other members of their species. Either that, or they're being manipulated by a parasite. In rare cases, intelligent animals like dogs or dolphins will drown or starve themselves, but only when they're confined in isolation and experiencing severe grief and depression. Humans are the only animals on earth that use weapons, poisons, or other active means to die by suicide, and we do it shockingly often.

Suicide is the tenth leading cause of death in the United States. In 2017, more than 50,000 people died by suicide, and 1.4 million people made an attempt. Middle-aged white men and LGBTQ teens are two of the most at-risk groups, a rare point of commonality between those two populations.

It makes sense that suicidal behavior is rare in the animal world. Animals are driven by a powerful set of instincts, honed by millions of years of evolution, to engage in behaviors that help them survive and reproduce. But so are we. That's what shocks me about the suicide epidemic in our species. How are we able to stare down our survival instincts and make a decision to die?

I believe language is the primary factor. Language, the collection of symbols and constructs that modern humans use to build a "map" of reality, has an unbelievable power to shape how we understand the world. Sometimes, this map of reality turns against us.

The most compelling theory I've read on suicide comes from Thomas Joiner, a psychologist and researcher at Florida State University. Dr. Joiner has his own tragic history with the phenomenon. In his book *Why People Die by Suicide,* he describes how his gregarious and popular father died by suicide while Joiner was in graduate school. I won't recount that story here—it's shocking and heartbreaking. But it's part of what prompted Joiner to write his book.

Dr. Joiner's theory is powerful in its simplicity. By examining thousands of deaths by suicide and interviewing hundreds of suicide survivors, he learned that there are three beliefs a person must hold to attempt death by suicide:

I am alone.
I am a burden on others.
I am not afraid of death.

Joiner's theory helps explain why many antidepressants actually increase suicide risk among those who take them—a fact you're likely aware of if you've ever listened to the lawyer-mandated voice-over at the end of a pharmaceutical ad. For most people, antidepressants offer relief from depression. But in many people, antidepressants have a reductive effect on all your feelings, including fear. So what happens when someone who already believes they are alone and a burden on others can no longer feel the natural, powerful fear of death we're born with? They slide into a state of suicidality. They may feel less depressed, but they still experience all three suicide factors.

This is the incredible power of language: Three simple beliefs, when internalized by a human being, can place that person in immediate risk of dying by suicide. Words do more than simply communicate, or even shape our experiences. As we saw earlier in the book, they have the power to create profound healing. But in some cases, words also have the power to end life.

I want to be clear about a couple of things before we continue. First, Dr. Joiner's theory is compelling, but it doesn't help to explain every suicide. Second, if you read *Why People Die by Suicide*, you won't actually find those exact three phrases in the book. Dr. Joiner describes his theory in much more clinical language. The list above is how I would distill his considerable work for those who are not in the habit of reading research papers for fun.

With those caveats aside, I want Dr. Joiner's theory to become common knowledge. Our culture has a powerful stigma against talking about suicide, in part because of religious beliefs that deem suicide an "unforgivable sin." If you feel alone and there's social pressure against talking about thoughts of suicide, you will feel even more isolated at the precise moment when you need to feel like you matter to other people.

The first time I encountered Dr. Joiner's theory, I couldn't stop crying. I knew exactly what the progression felt like, from feeling alone, to feeling like a burden, to finally being unafraid to die. As a survivor, I hear a constant "siren's call" from suicidal tendencies. Stressful days, times that I fail in my work and relationships—hell, even bad traffic jams and missed flights—can create a whisper in my ear: "You don't have to do this. You don't have to deal with it. There is an easier way out."

In the opening of the book, I said I never put shotguns in my mouth anymore. That's true. But I do live with suicidality as a constant presence in my life, and the reason I am still here is because I am not ashamed to talk about my struggle with suicide. I am not ashamed because my close friends and family have sent me signals, over and over in ways both subtle and overt, that I can talk about wanting to die and in that moment I will never feel alone or rejected or strange.

Last year, I told William, Michael, and Hillary, my fellow hosts on *The Liturgists Podcast,* that sometimes words of criticism lead me to fantasize about taking my own life, and the main barrier is not wanting to leave Jenny as a widow or my daughters without a father. I wish you could have seen how warm their faces were, their eyes brimming with tears, as they told me they loved me, that they needed me, and that they were happy to be my friends.

My friends have trained my brain to understand that I am not alone, and I am not a burden. My friends sing a louder and more beautiful song than the sirens of suicide island.

Isn't it strange that we can so often hear a voice in our heads say, "You are alone, you are a burden on those you love, and death would be better than living"? That accusing voice, ever present in our lives, can reach a climax that takes our suffering across a lethal threshold. But a few years ago I heard something that made me think of my inner accuser in a new light.

When I was just starting on my first book, I went to hear a man named Al Andrews give a talk in Nashville, Tennessee. Al is a therapist who specializes in helping creative people work through the complicated emotions that come up when you try to actually make something instead of just talking about it.

The people listening to the talk were all creative people: songwriters, poets, authors, and other people who work with words. I was an aspiring writer at the time, with dreams of talking in candid ways about the way people in faith transitions are often marginalized by the communities they rely on most. I knew I had something I wanted to say to the world, but I couldn't overcome the power of the inner accuser's voice.

I was perpetually frustrated as a writer. I had begun as a tech blogger, of all things, and enjoyed that well enough. It's easy to write what you like and don't like about a given product or service. But the kind of writing I wanted to do, about matters of spirituality and science, was much more difficult. Offering your opinion about a video card is one thing, but telling people how you coped with the loss of your faith felt far more unguarded (and dangerous).

Here's what that looked like for me. Anytime that I got an impulse to start writing, that inner accuser would whisper something in my ear. Sometimes the accuser would say,

"Who are you to write down your thoughts? No one will want to read them." If I pushed through that thought, the accuser would not relent: "Even if someone wanted to read what you write, you are a terrible, unoriginal, and boring writer." These thoughts were always much worse if I'd recently read something written by an author I admired. Most of the time, I didn't write at all.

Following the advice of a friend, I started a discipline of daily writing in 2013, but that didn't silence the accuser one bit. Every morning, before my family got out of bed, I'd walk into my home office and write at least a thousand words toward a blog post, or the notes that ultimately became my first book, *Finding God in the Waves*. I was painfully self-conscious about every word I wrote. And you know what? I was right. I really was a terrible writer. I've gone back and reread some of those early attempts to write a book, and they were absolutely awful.

But back to Al talking about the voices in our head. He told us about the times he'd heard those voices when he tried to create things. Al traced them back to childhood: times that he'd held up a picture or told a story and been criticized or shrugged off. I bet if you think hard enough, you can remember a time like that in your life. Some moment before cynicism set in, when you were able to make something and feel proud of it, but found that your joy wasn't mirrored in people you shared it with.

Like my inner-child self from chapter 4, who wakes up only when I'm in danger, the accuser isn't an enemy. That voice in our head is the voice of childhood scars trying to protect us from being hurt again by sharing our thoughts, our creativity, and even our feelings with those around us. The accuser comes from the parts of our social mammalian brain that equate fitting in with survival, with our reputation being a predictor of reproductive success.

Even as I write this paragraph, the accuser is there telling me that my thoughts are unoriginal, my prose weak, and my sentences too prone to land in the rule of three. If I let that voice win, then someone out there who struggles with suicidality as well never gets to hear they aren't alone when this book reaches them. But if I fight that voice, I've found that I get exhausted to the point where I can't write anything at all.

So I don't fight, and I don't surrender. Instead, I shift toward a posture of gratitude. I say to the accuser, "Thank you for trying to protect me. I know the pain you've felt—after all, I am you, and you are me. We are parts of a brain using language to shape signals across neural pathways. You are right, sharing my work is a risk. But not sharing carries risks too. Who is lonelier than one who says nothing at all?"

I do this before I walk onstage. I do this at dinner parties. I do it all the time because who I actually am is a sensitive and shy person who wants to be loved, and even more than that, wants others to feel loved. The accuser fears the exposure, and in response I do my best to hide nothing. What has the accuser to work with if I am someone who is honest about my feelings? The risk evaporates. I no longer have to wonder "What if?" Instead, I am freed to focus on "what is."

. . .

EARLIER IN THIS book, we talked about the power (and considerable brain mass) of the emotional parts of our brains. We also talked about the incredible power our feelings have in shaping our beliefs and behaviors. That led us to talk about "bottom-up" approaches to mental health like AEDP, the "triangle" of feelings from chapter 3. So many of us haven't been taught how to regulate our emotions in a healthy, adaptive way. That's what makes approaches like AEDP, with their emphasis on feelings that originate more in the lower structures of the brain, so helpful.

But that's not to say our cognitive faculties are unimportant. Both the experiment with blue-green discernment and Joiner's theory of suicide tell us that language has an incredible capacity to shape our experiences. Often, as with the accuser's voice, that "shaping" feels more like a binding. But it doesn't have to be that way. One of my favorite psychological tools happens to be a top-down, language-focused system for facilitating growth and change.

That system is called cognitive behavioral therapy. I've done a lot of it. It feels like you're reprogramming your brain with the same ease and elegance with which a programmer writes computer code.

Before I was old enough to drive, my mom found a notebook in my room. She wasn't snooping or anything. She was trying to clean up the mess that resulted from my combination of adolescent immaturity and depression.

That notebook was open to something I'd been writing about pain, and in large block letters I'd written:

SUICIDE IS THE OFF RAMP.

Today, as a parent, I have a much better idea of how it felt for my mom to read those words. It goes without saying that every parent's worst fear is the death of their child.

After a couple of difficult conversations, Mom made an appointment for me to see a therapist. A few days later, I was sitting in his office. I remember thinking that he looked like an actor playing a therapist. He had wispy gray hair and wore a plaid shirt and one of those blazers with leather patches on the arms.

We talked for a while, you know, like you do in therapy. I remember thinking that this was pleasant and all, but it didn't really make me feel any different. I still felt an overwhelming desire to die. But toward the end of our session, he handed me a little yellow book and said, "Michael, I'd like you to read this book and then tell me what you think about it when you come back next week."

"What is it about?"

"Well, it's about a kind of therapy where you use your thoughts to influence your feelings. The way you describe being aware of your thoughts as they happen makes me think this could be interesting for you."

So I took the book home and read it.

One of my biggest problems as an author is that most of the significant moments in my life have come from me reading other books. It's not a boring way to live—not at all—but it makes for boring reading if I write books about reading books. I mean, just imagine if I actually did that.

> *I was sitting at the desk in my bedroom, the only place that teenage me felt truly at home. The room was lit only by my desk lamp, but I was scarcely aware even of that fact. Instead, I was enraptured by each and every page of a small yellow book. It felt like reading the manual for the human brain, and it made me feel like learning the insights here would offer me the tools to leave my depression behind once and for all.*

Oh, that actually wasn't half bad. I surprised myself there. You're a reader, right? I mean you're reading this book. You get it. Books are awesome. And this little yellow book . . . wow, it changed my life. The funny thing is, I can't really remember anything about the book itself. No anecdotes, no research, nothing. All I remember is the system the book described: a four-step process for reprogramming your brain.

Identify a problem. In the case of my adolescence, that was "I am suicidal and depressed because I believe I am unlikable."

Set a goal in response to that problem. For teenage Mike, this was "I want to be happy and have friendships."

Pay attention to your thoughts. Don't just let them happen; actually listen to them.

When you have thoughts that discourage you or make you feel bad, interrupt and correct them. So if you notice yourself thinking, "No one wants to be my friend," stop your train of thought and push back by thinking (or even saying aloud), "I am likable. I know that because Jon, Ketan, and Todd are my friends."

As you'll see in the rest of this chapter, that's a wildly oversimplified description of CBT. I mean, my favorite book about CBT these days is called *Cognitive Behavior Therapy: Basics and Beyond,* and it clocks in at 391 pages. The little yellow book I read (the name of which is lost to me) was still probably 130 pages.

By the time I went back for my second therapy session, I was really excited. Paying attention to my own thoughts had never occurred to me before. How can someone not notice their own thoughts? But once I tried observing my own thoughts, I felt somehow separate from them—and that distance made me feel less helpless in the face of my antagonistic inner voice.

If you think about this in the context of my "person standing on a dog standing on a crocodile" model from chapter 2, CBT is about the person training the dog and the crocodile instead of trying to outmuscle them. It's a way of confronting the automatic responses we have to most situations, which often goes like this:

Something happens → We have "automatic thoughts" in response → We react to the something that happened with our feelings, thoughts, and behaviors.

This cycle of automatic thoughts and reactions can lead to damaging core beliefs about ourselves. When I was a teen, I had a core belief that I was unlikable. That belief was formed by real experiences: I was an unpopular, bullied child with learning and developmental disabilities. So even as puberty and adolescence changed my body, and the less restrictive social hierarchies of high school allowed me to make numerous friendships, that old software code was still running at the base of my thoughts: *No one wants to be my friend.*

This led me to discount any cues that undercut the "I am unlikable" story. Teachers often told me that I was a wonderful student and a kind person, but I would discount those moments as obligatory statements from authority figures. Likewise, the consistent affection my parents offered was dismissed as parental instincts that prevented my folks from seeing the real me. Even as I became more popular at school—to the point that people by the hundreds would come to see my band play—that new information would deflect off my core belief without altering my perception of self at all.

Hence the metaphor of CBT as a way of programming your own brain. You stop "running" the program of automatic thoughts and reactions, and you start "writing" a new one with intention.

So if you want to exercise but struggle to do so, CBT posits that you are stuck running old code. You think about exercising but then have an automatic thought in response. Perhaps "I'm too tired already," or "the weather isn't right," or even, "I can just exercise tomorrow." But that creates a feeling of disempowerment and dysphoria that further drains your energy and leads to a behavioral outcome: staying in bed.

In order to break that cycle, CBT recommends that you first become aware of the pattern and then confront it. When you think "I'd like to exercise" but then automatically think "I'm too tired," you challenge the assumption in your mind. "Am I really too tired? So tired I can't even move around the room? In fact, I could move around the room right now." When you get up, you will often find that you feel better, having done *something,* and you may even go for a walk, since that will help you sleep better later.

The repetition of the observe-interrupt cycle is how CBT powers change in our behaviors and reforms old patterns of automatic thinking.

CBT IS MEANT to be a time-limited form of therapy. When I encountered it as a teenager, I was slated to do six sessions, but I felt much better by the time I'd done three. Don't hear me wrong: CBT isn't a miracle cure. Around 60 percent of patients see improvement in symptoms of depression during the course of CBT. That means around 40 percent of patients don't see improvement. I'm not selling you a miracle cure to your problems.

Instead, the reason I highlight CBT in this book (in addition to bottom-up approaches like AEDP) is because it helps explain why self-help fails to deliver lasting change for so many people. There is no one-size-fits-all approach to improving mental health. Quite the contrary—there is no single approach that will work for even *one* person all the time. CBT helped me escape suicidality as a teenager and empowered me to face countless challenges into adulthood. But AEDP was much more effective in helping me deal with panic attacks and anxiety.

Our brains are remarkably complex and intricate. And so are we. Your brain-body systems have given you the ability to adapt to so many situations, but the very complexity that allows you to thrive makes troubleshooting your own problems more difficult than rebuilding your car's engine.

In the miracle of you, words have incredible potential to build you up or tear you down. Which is why my favorite part of the entire Harry Potter series is this quote from Albus Dumbledore: "Words are, in my not-so-humble opinion, our most inexhaustible source of magic. Capable of both inflicting injury, and remedying it."

Dr. Joiner's theory of suicide and cognitive behavioral therapy tells me the person standing on the dog standing on the crocodile has her own strength. When we stop fighting our emotional brain, and instead learn to train and trust it, we have the power to confront, influence, and change the words that run through our own mind in a way that can empower us, heal us, and help us be the people that we want to be.

A Box Full of Mousetraps

The Delicate Dance of
Living Among Humans

One of my favorite pastimes is studying the endless, different hypotheses about where consciousness comes from. No one really knows why we experience our own thoughts as a story, or have self-awareness, but I love what the famous physicist Michio Kaku has to say in his book *The Future of the Mind.*

Here's the basic idea. Consciousness is any feedback loop that models an environment, can impact that environment somehow, and responds to the consequences of those changes. Don't worry, I didn't get it at first either.

Let's start with the most basic kind of consciousness Kaku describes: the humble thermostat. (If the idea of a thermostat being conscious raises strange questions about a Honeywell's right to life, stick with me for a minute.) A thermostat builds a model of reality. It's a simple model—all your thermostat understands about the world is the temperature—but it's a model nonetheless. A rock, by contrast, is completely passive. It doesn't build any model of reality; it just is.

Thermostats are aware and active. They know the temperature *and* they can do something about it. If the temperature is 76°F, a thermostat will respond by closing a circuit and activating the air-conditioning system. It keeps that circuit closed until the temperature falls below the set value, say 74°, at which point it will open the circuit and turn off the air-conditioning. In Kaku's model, this simple single feedback loop forms a basic consciousness, or what he calls a Level 0 consciousness.

The consciousness of a living organism is more complex. Consider a sunflower. Sunflowers have many more feedback loops, and they build a more elaborate model of reality. A sunflower is aware of the temperature, the humidity, the direction of the sun, the pH and moisture of the soil, and the presence of parasites. Sunflowers can respond to all those variables. Sunflowers, with their dozens of loops, form a higher Level 0 consciousness in Kaku's model.

The next step of sophistication is a Level 1 consciousness. Think of a beetle. Beetles have to build an even more complex map of the world than sunflowers do. They have to know where they are, and what's around them. They have to know the difference between things they can eat and things that can eat them. They have to find other beetles to make more beetles. It's quite the grind compared to life as a plant, so beetles need hundreds of feedback loops, many of them centralized in a brain, to do what they do.

Jump from hundreds of loops to thousands, and you'll get social mammals (Level 2 consciousness). Social mammals have to know everything a beetle knows, but they also need to understand where they stand among members of their community. To do that, they have to imagine what other individuals in their community feel about them. Wolves are a great example. Like all canines, they have remarkably sophisticated social and emotional intelligence, and their model of reality has to incorporate the models of reality in the brains of *other* wolves. That's pretty insane.

Kaku puts humans alone in the third level of consciousness, mostly because they have a unique set of feedback loops devoted to understanding time. Humans have an unmatched capacity to dwell on the past and scheme about the future, and we need very large brains to do so.

Still, like wolves, you keep little replicas of other animals' consciousness in your own map of reality. These models are quite good—people are remarkably adept at imagining what others think and feel about them. The people we know best have the highest-resolution "models" of us, which is why you probably know what your mom would say if she saw you got a tattoo.

My point here is that our brains didn't get like this so you could do physics or build computers. Most of the reason you have such a big brain is so you can fit into the complex culture created by your fellow social mammals. Evolution has been driving our branch of the tree of life in the direction of social collaboration—and competition—

since dinosaurs ruled the earth. As far as we know, we're better at doing this than any other creature.

So why is living among other people so damn hard?

I GREW UP in the church. Our congregation was a place where no one made fun of me, and a casserole was almost always within reach. But I can remember times when the relative calm of church life was punctuated by periods of unhealthy conflict. Factions could form in the congregation, divided over matters as simple as the style of music or the color of the carpet, or more substantive matters, like how we should help people find relationships with God, or meet the needs of the poor around us.

Often these disagreements would lead to power struggles—and in one particularly egregious example, the church I grew up in actually split. Almost half of the members left to start a new church altogether. Those times of conflict raised fundamental questions about how I saw the world. I thought the church was meant to be a place that introduced people to God. Who cares if we have drums or shag carpet? Millions of people are headed for Hell if we don't let them know the truth of Christ—or so I believed at the time.

Of course, secular communities can be just as warped as faith-based communities. Every human institution, from a multinational NGO to a tight-knit group of friends, is prone to this same cycle, in which a community begins with symbiotic joy before sliding into mission drift, power struggles, backbiting, and, ultimately, a split or dissolution. The problem with my childhood church wasn't the church at all. It was that the church was made up of people.

Think of the personal stories I've told across the arc of this book. I've experienced considerable pain, and I don't always have the best strategies for coping with it. When someone does something that pricks a past wound, intentionally or not, I won't necessarily respond to the person in front of me. Instead, I'll react in a way that protects against a past wound. We all do this.

But the cycle doesn't stop there. As I react to protect my wound, my reaction may trigger a wound of your own. That's complex enough in a two-person conversation, but what if it's a family gathering? Now dozens of people are reacting and becoming dysregulated in a way that ratchets up the emotional intensity of the moment, uprooting even more pain, wounds, and traumas.

Zooming out to the scale of a company, church, or charitable organization, you can move from dozens of people to hundreds—even thousands. At this scale, not everyone has relational intimacy to use for conflict resolution, so empathy falls even lower, the stakes feel higher, and people fight harder. It's like dropping a ping-pong ball into a box full of mousetraps. One small action produces mass chaos.

Most of this book has been about you learning to love and accept yourself as an individual. But when "you" becomes a plural pronoun, the hidden forces that shape us grow exponentially more complex. It's hard enough to manage our own shadows, much less a room full of them.

. . .

BACK IN THE 1960s, a psychologist named John Bowlby was studying why children have different temperaments. He found that a child's disposition was strongly correlated with the presence (or absence) of a strong attachment with at least one caregiver. Children with a strong attachment to a caregiver were more outgoing, calm, and resilient than those who lacked a strong caregiver attachment. In the intervening decades, Bowlby's observations have grown into an entire field of psychology called attachment theory. In my eyes, it helps us understand relationships in the way that knowing the earth is a sphere helps us do geography.

As more psychologists contributed to attachment theory, four stages of the attachment process emerged from research.

Pre-attachment stage (birth to six weeks). During pre-attachment, babies don't show any preference for a particular caregiver. They cry when they want or need something, and settle down when anyone meets those needs.

Indiscriminate attachment stage (six weeks to seven months). Babies in this age range will still respond to care from anyone, but they begin to form a preference for a primary caregiver. This is why babies start to fuss more when someone other than "Mom" tries to comfort them.

Discriminate attachment stage (beginning around seven months). As anyone who has ever held a baby in this age range can attest, babies in this stage have a strong preference for one particular caregiver. They get angry or afraid when separated from this person, giving rise to anxiety in the child's life. The child also becomes fearful of people they don't know (called stranger anxiety).

Multiple attachment stage (beginning around ten months). I was elated when my daughters hit the multiple attachment stage because suddenly they seemed to like me. Prior to this, they were fine if, and only if, Jenny was around. But once children reach the multiple attachment stage, they still have a primary caregiver they prefer, but they form independent bonds with other caregivers—to the joy of grandparents everywhere.

These stages are based on a critical assumption: that the infant has a primary caregiver who will respond to their needs in an appropriate and supportive way. But as researchers studied attachment in the field's early years, they found in data what we all know through life experience: Not all caregivers respond in a consistent and supportive manner to the needs of young children. When this happens, babies don't have the context to cope with maladaptive behaviors from the people they trust most. They don't know how to separate real trauma from a bad night's sleep. All an infant's brain understands is if their care is consistent and supportive, or not.

Let's pause and consider something here. Attachment theory says that much of the way people approach relationships is shaped in the *first year of life*. I don't know about you, but I don't remember anything before my third birthday, much less my first. But the research says I can know a lot about what life was like for me as a baby based on how I behave in relationships now.

That's the power of attachment theory. Not only does it give us a glimpse of our life beyond what we can remember, but it also offers tremendous insights on how we behave in long-term relationships with others based on our attachment style. There are four major attachment styles.

The first, and most common, is **secure attachment.** This is the most common attachment style, meaning at least one of your caregivers was consistent and supportive in how they cared for you. People with secure attachment tend to be confident in relationships and self-assured. They feel comfortable asking for support when they need it, and just as easily offer support when their partner needs it.

People with **anxious attachment** had parents who were supportive sometimes, but not consistent. They are desperate to make a lasting bond with another person, so much so that they often engage in behaviors that drive people away. They can be clingy, jealous, or demanding when they feel insecure, and may even "test" a relationship by creating conflict. People with anxious attachment are looking for someone to complete them.

The last two attachment styles are "avoidant" types. Both avoidant types were shaped by a caregiver who responded with withdrawal, anger, or another punitive reaction when they expressed their needs as an infant.

Dismissive-avoidant attachment looks like someone who doesn't need other people. This is an illusion, of course. All humans need intimacy with others in order to be stable. People with this attachment style learned to avoid showing needs to their caregiver based on how their caregiver reacted when they expressed their needs. As adults, they launch preemptive strikes against rejection by hiding their need for connection with the people they love. They appear detached or aloof in relationships, and they avoid relying or depending on others. In times of conflict, dismissive-avoidant types have the capacity to shut down their emotions and respond with indifference if a relationship seems on the verge of irreparable harm.

On the other hand, **fearful-avoidant** types know they want to be loved but are terrified of being close enough to others to be hurt. People with this attachment style tend to have unpredictable and explosive emotional displays in their significant relationships. They're in a constant "push-pull" cycle trying to draw others near enough to meet their needs while keeping them at arm's length as a defense. Relationships with fearful-avoidant types are roller-coaster thrill rides with soaring highs and crushing lows. These types lean into their partner for safety, but start feeling trapped once they get close enough to feel safe.

Research tells us around 60 percent of people are secure types. Another 20 percent are anxious, and the remaining 20 percent avoidant. As you can imagine, relationships between people with secure attachment styles tend to be the most successful—success being defined as a lasting relationship where both partners feel satisfied. What may surprise you is that secure types don't mix well with avoidant types, statistically, and that anxious types do better with other anxious types than they do with secure types.

To make matters more complex, you can have a mix of attachment styles, and you can have different attachment styles with different people. Our brains are sophisticated enough to create a mix of strategies based on experiences with different caregivers.

I've talked about attachment theory enough to know that people often worry when they identify as having something other than a secure attachment style. But remember: The relational "wiring" in our brains is set up when we're just babies! Your attachment style isn't a "life sentence." You didn't do anything to shape it, but understanding your attachment style can free you from maladaptive behaviors in relationships. By doing attachment work, generally in conjunction with a trained specialist, people of any attachment style can move toward an "earned secure attachment style," or as I call it, "Undoing What Your Parents Did to You."

A FEW MONTHS into working with him, I told Ron that I find it hard to reach out to the people I care about. I mentioned that Hillary had called me the night before and done something I would rarely do. Partway through our conversation, Hillary told me I mattered to her and that our relationship was really important to her. I froze. I didn't know how to respond.

Ron asked me, "How does that make you feel?"

When people ask me how I feel, I almost always respond with a sentence that begins with "I feel like . . ." But Ron doesn't let me offer "I feel like" responses; he says those are thoughts, not feelings. So I did what Ron has taught me to do. I sat still and waited to see what I felt in my body. After a few moments, an answer took shape.

"I don't feel anything at all," I said. "Not warmth, not tightness, not anything. I never believe it when someone tells me that I matter to them."

Ron smiled and said, "That's good."

"Wait, what? That's good?" Was this a trick question?

"No, Mike, it's good that you told me. I know it can be hard for you to share things like that."

Then Ron asked me, "What's behind that?" And I said, "I don't know."

Ron's expression was warm and open in response to my pensive posture.

"Let's just sit with the idea for a minute. I want you to imagine Hillary's face and imagine that she likes you. Just sit with that and tell me what happens in your body."

So, I did.

ALMOST EVERY TIME I listen to my body, it says, "Let's go get a pizza." But not this time. As I imagined Hillary's face and imagined that she liked me, and valued me, and was telling the truth when she said she was grateful to be my friend, my body responded. I felt a pressure in my chest and heat in my belly. Ron said to keep waiting, because it meant feelings were on the way.

Then a dam broke, and I started to sob. I felt such grief, sadness, shame, and fear. I knew that if Hillary really knew me—not the veneer I show the world, but who I really am—she would never want to talk to me again. If Hillary knew I was still the same shy, scared, overweight child inside, why would someone so talented want to waste time being my friend? I felt shame over who I am, and terror at the idea of Hillary ever finding out.

In a break between sobs, I told Ron that none of this makes sense: "How can I be so sad that someone I love and respect like Hillary likes me?"

"Don't worry, Mike," Ron said. "We'll talk about that in a minute. For now, just pay attention to what you feel." That was the moment I knew that Ron was a good therapist for me—he knew my brain was a storytelling machine. In order to be present now, I needed the promise that I could explore an explanation later.

As I wept, something remarkable happened: I remembered my childhood. That never happens to me. Prior to writing this book (and going to therapy with Ron), most of my life before high school was cloaked by dense fog in my memory. But as I sat in Ron's office and thought of Hillary, I suddenly remembered my kindergarten classroom with stunning clarity, from the shape of the room to the placement of the chairs, all the way down to the

cartoon alphabet on the windows that faced the playground.

I remembered my teacher's face, etched with concern
as she watched my classmates tease me with such ferocity
that she couldn't maintain order in the class. I was a fat,
little autistic kid, and the way I interacted with other children was a magnet for mockery and scorn. I remembered
how powerless and defeated she looked as she led me out
of my seat and over to the time-out chair, to separate me
from my classmates.

Then I remembered her telling the class they were all in
time-out, and telling me that I was not. But the end effect
was the same—I sat alone while all of the other kids sat
together.

As I recalled that story, I told Ron I didn't know what
the memory had to do with Hillary. Ron said, "Based on
all you've told me in our sessions together, what happened
that day in kindergarten was not an isolated incident.
Throughout your childhood, you were rebuffed when you
tried to connect with others, and your nervous system re-
members. You often don't feel safe in relationships now,
because you were not safe in relationships as a child. Your
nervous system views reaching out as dangerous, and your
brain has stored that as trauma."

This is why I am a forty-year-old man who never texts
or calls people. Not my friends. Not my family. Not my
wife or children. Whenever I try to do so, my brain offers
me a sense of unease, fear, or even panic. The dog drops its
tail, while the crocodile bares its teeth and hisses. It's safest

to sit in time-out, alone with my thoughts and imagination.

But people grow when they are loved well, and I am loved very well by the people around me. I am fortunate enough to have people who push past my dismissive-avoidant patterns. Of course, pressuring someone in a dismissive-avoidant frame is *the worst* way to get close to them. What I'm talking about is people who are persistently present, but noncoercive. People who love me without an agenda.

My friend and cohost Michael, a famously introverted songwriter, not only pushes past my reticence to reach out, but delights in every strange, quirky, or bizarre part of my personality. His constant and unconditional acceptance helped me grow comfortable being "me" for the first time in my life—an arc you can hear if you listen to the first few seasons of our podcast.

My friend Micky ScottBey Jones, an activist, organizer, and speaker whose work I admired long before I had the opportunity to be her friend, challenges me to show up with her in friendship—not just in her advocacy or public work. She's always less interested in how I can promote her work than in sitting together for coffee whenever we are lucky enough to be in the same city.

Then there's the group of friends in Los Angeles that I spend most Sunday nights with. They've shown up in times of crisis, of course, but even more remarkable is the way they let me move in and out of our weekly conversations, balancing my desire to participate with the over-stimulation autistic people often feel when they are among groups of people.

The steady and gentle manner in which my friends reach out to me is the very thing that allows me to reach out to you in podcasts, in appearances, and indeed, in this very book.

Friend, whatever your attachment style, whatever wounds you carry, whatever trauma has shaped your growth—it is not a liability. The pain I've endured has made me incredibly sensitive. That sensitivity can be debilitating, but with support from others, that same sensitivity has become one of my greatest strengths.

One of the main challenges of life is learning to support others from our strength, and accept the support of others in our weakness. There's no shortcut to this process, and some of us begin with significant deficits. But humans are social primates, and I don't believe there is any path to health and wholeness that can be taken alone.

The Forest and the Trees

*Why Learning to Live in Our Bodies
Might Save the World*

If you search YouTube for "white people dancing," you'll find videos of middle-aged Caucasians doing hilarious, awkward dance moves. When I share these videos in presentations, people laugh until their eyes water. Not me, though. I only wish I could have moves like a white person with bad rhythm.

One of my favorite things we do with The Liturgists is host dance parties for people who support our podcast. We'll rent a venue that serves booze, play some loud music with a beat, and dance the night away with a few hundred folks from that city. It's great fun, especially since so many people at the party (the hosts included) grew up believing that alcohol was liquefied sin put on the earth by Satan to destroy people's lives.

Our parties didn't start as dance parties. When we started throwing them, we mainly meant for people to stand around and talk. But the people who listen to our podcast are younger, and most of them are women, so at each party the background music got louder and louder until it wasn't background music at all, and then people would start dancing.

Not me, though. I don't dance.

I'd like to say it's shame or fear that holds me back, but it isn't. There's some fundamental circuit between my brain and my hips that isn't connected. I have enough trouble walking without running into furniture and door frames. Dancing requires such intricate and complex body movements that I can barely follow it with my eyes, much less mimic other people's movements with my body.

Here's what I look like when I try to "participate":

Me standing still.

But then the music really gets going, and people start moving all around and I start to feel a tingle or something in my elbow, like something exciting is coming. Then I do this:

Me, still standing still.

Dance parties, embodiment exercises, gym classes—they're all the same. They all try to get me to do something that I can't do. Invariably, someone will see my distress and say, "Hey, don't worry about it. You're just left-brained." What they mean is, "You're a logical, awkward nerd."

I know they mean well, but it drives me crazy when people say that. Not because I'm offended, but because that's not what it means to be "left-brained." The real definition is much more consequential, a fundamental division in our brains that threatens our capacity to form a compassionate and humane civilization together. From our politics to our spiritual beliefs to our inability to hear the pain of others because we are too busy hiding our own, our inability to face some of the most decisive issues in society is born from the way we've emphasized the left brain at the expense of the right.

ALMOST EVERYONE KNOWS that our brains are divided into two hemispheres. As soon as we're taught this fact, we're also taught that the right side of the brain is creative, while the left side is analytical. I've heard this fact referenced in countless TED talks and textbooks. I've seen numerous illustrations and articles based on the premise. But none of that changes the fact that it's wrong. The idea that the right brain is creative and the left analytical is, ahem, dated in neuroscience. But it's simple and accessible, while also feeling applicable to life. (I'm left-brained! I'm right-brained! Yay!) Any good marketer will tell you that simple and accessible beats accurate every time.

Here's a more accurate idea: The right brain thinks holistically, while the left is more reductionist. Your left brain looks at a book and notices the paper, binding, cover art, and type. The right brain sees a *book*, or maybe even an opportunity to learn or go on an adventure of the imagination.

In other words, the right brain sees a forest, while the left brain sees trees. The approaches of the two hemispheres are complementary, and our brain shifts between strategies as needed. (By the way, both rational analysis and artistic creativity require a choreographed dance between the hemispheres. Just to lay that myth to rest.)

For most of the time humanity has existed on this planet, the two hemispheres of the brain existed in harmony. They helped us build tools, find meaning in the stars, coordinate complex hunting strategies, and tell amazing stories afterward about the hunt. The two hemispheres created language, culture, art, agriculture, and civilization together.

But it couldn't last. In the late seventeenth century, the left brain launched an all-out war on its counterpart, starting in Europe and what would become the United States. Today, this war is known as the Enlightenment, a movement of philosophers and scientists who set the cultural expectation that analytical, left-brain thinking was best. I mean, could there be a better slogan for how our left brain views the world than Descartes's "I think, therefore I am"? Or Immanuel Kant's statement "All our knowledge begins with the senses, proceeds then to the understanding, and ends with reason. There is nothing higher than reason."

The Enlightenment wasn't the beginning of the left brain's uprising—indeed, reason had already been tipping the scales of human experience in its favor in ancient Greece and China. But the Enlightenment took this war on right hemispheres and launched it on a global scale. The Enlightenment led to rationalism, modern science, and an ever-increasing suspicion of cultural traditions and emotions.

In the West, our socioeconomic systems were codified around this value of favoring logic and reason over other ways of knowing, like intuition and experience. Capitalism is a wildly left-brain way of running an economy. Only a reductive economic view could prioritize "shareholder value" every three months over the long-term viability of our soil, or the climate that allows economies to exist in the first place.

As a consequence, brain-imaging studies show us that modern Western people have a considerable bias toward using the left hemisphere of the brain when compared to other populations. Post-Enlightenment culture has rewired our brains, leading to incredible advances in science and technology. The computer I'm writing this book on is a child of Enlightenment-style thinking. As are antibiotics. But these advances also came with the cost of isolating people from each other and from the ecosystems that make our lives possible. What good is it for a man to have strong quarterly earnings but forfeit the ocean? We've decimated fish populations to the point where old favorites like Atlantic cod or bluefin tuna are disappearing from menus all over the world.

In many ways, the United States of America is the ultimate example of a left-brain culture. Our culture and our laws speak of the essential nature of individual liberty and self-sufficiency. These notions are fundamental to not only how Americans see society, but how they understand themselves. Like many Americans, I am naturally generous but reticent to ask for help. I prefer to pull myself up by my own bootstraps while imagining the smiling visage of Teddy Roosevelt looking down on me with approval.

That portrait of Teddy has reshaped my brain, and probably yours as well. The way we see the world actually impacts the ways our brains process information—and from there, how we live. Brain-imaging studies have shown that the left and right brains have oppositional approaches to integrating information. The left brain is dominated by "top-down" processing, in which the neocortex takes the onus to coordinate—and dominate—activity. Thinking back to the burrito brain, the left hemisphere begins more processes in the tortilla, before working its way down to the meat in the middle.

The right brain, on the other hand, is more focused on a "bottom-up" approach, allowing the instinctual and emotional centers of the brain to integrate into all forms of processing. That approach means that our right brains excel at relating to other people and understanding how our actions impact our world.

When we allow our pattern of cognition to cater to the left brain, we lose access to all of the richness that our brains are capable of offering. We can't see the forest for all the trees. And when you combine left-brain bias with

America's radical individualism, it's even worse. Humans are social primates. Like trees, we are wildly interdependent for not just survival but the quality of life. But our left-brain world tells us to see ourselves as individuals, creating disconnection and disembodiment as sad as a lone tree in the middle of a field, waiting for a storm strong enough to knock it down.

This self-imposed sense of isolation often leads us to hide our pain and difficulties in the very moments when sharing may be most likely to elicit support and solidarity from others. And that fear is born from a worldview that trains our brains to take everything apart.

HAVE YOU EVER wondered why some people live "in their heads," lost in an inner landscape of abstractions and thoughts, while others live with their toes in the sand, and an arm around a close friend? That's because left vs. right isn't the only interesting division in our brains. Though the lines of demarcation are less stark, there are also contrasts between the front and back areas of the brain—especially in the neocortex. That division has caused neuroscientists to argue about where consciousness happens in the brain.

The front-of-brain team advocates the global neuronal workspace (GNW) model of consciousness. Our conscious experiences incorporate relatively little of what happens in our brains. You're not aware of your brain correcting for the blood vessels in front of your retinas as you read this book, for example. The action is happening, of course, but it's operating outside your conscious experience.

The GNW model defines consciousness as whatever information makes it into our working memory—the network of brain structures that includes the prefrontal cortex, anterior temporal lobe, inferior parietal lobe, and other, less famous brain structures. Anything that makes it into this region of the brain becomes part of your conscious experience, and anything that does not remains subconscious.

In the other corner, the back-of-brain crew has advocated for a model called integrated information theory. The IIT model hypothesizes that our conscious experiences arise, not in our working memory, but in the parts of our brain where nerve signals become sensations. It's the difference between the automatic reflex of a jellyfish and a human feeling the gentle caress of a lover—something that we not only feel, but *experience* in a deeper way. These regions of the brain are called the neural correlates of consciousness (NCC), and are located in the temporal, parietal, and occipital regions of the neocortex.

Scientists discovered the NCC by zapping people's brains with electricity during brain surgery. I'm completely serious. Our brains are so individualized, thanks to neuroplasticity, that neurosurgeons can't map them in the way other organs are mapped. The part of your brain that helps you taste the flavor of pineapple, for example, may be where someone else has perfect pitch. This is why doctors usually keep patients awake during brain surgery, stimulating the brain with electrodes and asking the patient what they feel. When you zap the NCC, people experience sensations: being touched on the arm, tasting a strawberry, hearing music, or helping. It helps the surgeon avoid cutting a part of the brain that's essential to someone's quality of life.

But guess what happens when we zap the portions of the brain in the GNW model? Not a damn thing.

Armed with this finding, Team IIT argues that sensation is the foundation of consciousness, while the awareness facilitated by the GNW is simply a tool employed by our consciousness. The frontal lobe may be our narrator, our CEO, or our zookeeper, but the sensations that actually facilitate consciousness occur in the posterior regions of the neocortex.

The precise nature of human consciousness is an unsolved problem in science. Both GNW and IIT have evidence to support their theories. Notably, everyone on both sides of the debate is far more qualified to have an opinion than I am—after all, I'm a community college dropout. But I think it's appropriate to quote the wisdom of a fellow nonscientist, Forrest Gump, here: "Maybe both is happening at the same time."

It seems to me that the play between sensation and agency is precisely what it feels like for me to feel conscious. Three years ago, I would have sided with the GNW view of consciousness without hesitation. But, as I've learned more about the brain and gotten more in touch with my body and my feelings, I've become more aware of the limitations of left-brained human agency.

I've started to view the prefrontal cortex less as the brain's boss, and more like the brain's narrator. I think most of the time, our consciousness is a beautiful story told by the voice in our head, and that story is about the sensations and experiences we face every day. Even when the prefrontal cortex "takes charge," I think it works less like Jack Welch and more like Angela Merkel. A head of state in a modern democracy isn't an autocrat. She has to build consensus among and explain the actions of a chaotic cast of bureaucrats and constituents.

Either way, the prefrontal cortex is never the exclusive seat of our conscious experiences. Biologically speaking, humans can't live anywhere but in our bodies. But we can absolutely shift our awareness into a greater emphasis on cognition or embodied sensations, and that posture influences what parts of our brains play a primary role in our conscious experiences.

FOR THE LAST forty years, I've been a champion of Enlightenment-style thinking. My nickname is Science Mike, for God's sake. I've lived in my head, which is to say in a disembodied way, my entire life. I've been cheered on for this—my capacity to isolate my ideas from my identity and from my emotions helps me to navigate white spaces, male spaces, straight spaces, and Evangelical spaces like a native citizen. Which, of course, I am. I can turn any issue into a set of ideas that are open to identification, classification, and debate. As I do so, I can appear both "reasonable" and "calm." By participating in the world in this way, I become a "nice person," who moves through the world in a polite, civil manner. I take out the trash, pay my taxes, and ask "Did he do what the officer said to do?" in response to news of a police shooting.

I grew up in the years following the civil rights movement—just far enough in its shadow for the figures behind it to be sanitized in a way that helped a generation of white children believe that America had become a postracial utopia. Corporations had started hiring women CEOs, and my classrooms were full of people of all kinds of races and ethnicities. We all coexisted, and being civil toward others was a primary value. It seemed like America was close to reclaiming its founding promise: a nation where all people were created equal, under God. That is a very easy thing to believe if you live in your head, but not so much if you live in your body.

I have learned that the niceness that makes my life easy also makes the lives of my friends very difficult. Which friends? The ones who are not white, straight, or Christian. The same set of social codes that allows me to thrive prevents my friends of color—let alone my friends who are women, queer, or disabled—from describing their life experiences to others. The niceness that is so often seen as the glue of our culture, and the disembodied manner in which white people approach our lives, means the voices are dismissed if they are "too emotional" or "mean." If a black friend of mine gets angry because her predominantly white workplace won't discuss—or even tolerate her discussing—the presence of bias in the company's hiring practices, she is twice marginalized. Once for her race, and a second time for her anger.

In this way, "niceness," which is to say, distancing our-selves from uncomfortable feelings, becomes a veneer that masks the cruel and inequitable foundation of our civiliza-tion. It encourages us to police any behaviors and feelings that provoke discomfort, and in doing so, we oppress and disenfranchise people, sometimes without intending to do so.

The tragedy is that this state of affairs benefits no one. There are economic benefits to being white, to being a white man especially, but that gain is offset by severe costs to mental health and social connection. As we discussed earlier in the book, white men are so lonely that they are killing themselves in record numbers. The individualistic framing of our culture, created and maintained by men of European ancestry, has created a box that simultaneously elevates us in society while slowly killing us.

In one recent study conducted by researchers at Unilever,* this confining set of ideas was labeled the Man Box.

THE MAN BOX is a set of enforced behavioral norms for men. Researchers discovered it by performing in-depth in-terviews, psychological evaluations, group discussions, and other tests in a random sample of men ages eighteen to thirty in the United States, the United Kingdom, and Mexico. I like to think of these norms as a series of Seven

---

* Funded by Axe—yeah, the body spray people. The truth is stranger than fiction.

Commandments, except instead of being written on stone tablets by God and delivered by Moses, the Man Box came down the mountain on the tailgate of a pickup truck.

Here are the Seven Commandments of the Man Box:

- THOU SHALT BE SELF-SUFFICIENT.
- THOU SHALT ACT TOUGH AT ALL TIMES.
- THOU SHALT BE ATTRACTIVE, BUT THOU MUST NOT WORKEST TOO HARD ON IT.
- THOU SHALT ONLY DO MANLY THINGS, FOR COOKING AND CLEANING ARE WOMEN'S WORK.
- THOU SHALT NOT BE GAY.
- THOU SHALT THINK OF AND PURSUE SEX AT ALL TIMES.
- THOU SHALT TAKE CONTROL OF YOURSELF AND OTHERS THROUGH KICKING ASS.

These commandments tell men that they have to rely on themselves, and that vulnerability is a weakness they can't indulge. There are penalties for nonconformity—my childhood is a testament to that. Any man who fails to live within the box will face everything from subtle shunning to outright verbal and physical violence.

The self-sufficiency of the Man Box is completely antithetical to real relationships, of course. It means men can't share their feelings, even feelings of loneliness as their life progresses, and the ones they have are often confined to superficiality. The demand to "act tough" means the only feelings that men are allowed to express are joyful frivolity and thrill-seeking, on one hand, and anger on the other. To be "tough" is to be as one-dimensional as an action movie hero.

The Man Box offers conflicting orders when it comes to appearance. Men should be attractive, mainly through working out, but shouldn't put too much effort into their hair and clothing. After all, to fuss over one's hair and clothing would be "gay," and nothing is less manly in the Man Box than gayness.

The Man Box says men should make money, but not worry about taking care of the home. This assumption is causing a generation of men to feel lost as women begin to gain income parity with their male counterparts. It also means women who share the burden of income generation still do a wildly disproportionate share of the housework and parenting.

The Man Box tells men to be assertive, to be aggressive, to dominate, and to win. It tells them to take charge and force their will on others in every situation. Men who do this are rewarded with promotions, prominent social standing, and admiration.

I've never measured up to the Man Box—but I have tried. I have faked self-sufficiency to the detriment of my adult life. (In the next chapter, you'll see how deep that rabbit hole goes.) I've felt incredible shame as I've grown older, become less obsessed with sex, and stopped viewing women's bodies as objects of conquest. The Man Box makes me angry because my failure to "measure up" as a sensitive and nerdy man has always led to pain and rejection when I am with other men.

My disinterest in sports, rough physical play, and crude jokes about women has often landed me with scorn, or outright questions about my sexuality.

But I'm learning that I've been lucky that way. As my peers have gotten older, I've noticed that the ones who were most successful at living in the Man Box are now, in many cases, the most lonely and broken people that I know. What does a man do when the Man Box told you to be successful but all that focus on work made your marriage fall apart? What good are a salary and stock options when you don't have any friends?

The Man Box keeps men from having honest conversations about their feelings, their fears, or their insecurities. How many men crave intimacy but can't take the first step because they are afraid of appearing weak? How many men have fallen into cycles of shame and despair over erectile dysfunction because they haven't been given the tools to explore intimacy in conversation, let alone in sex with their partners?

How many women are trapped because they've been told they can't be assertive without being a "bitch"? How many gay men have felt they aren't really men because they don't have sex with women? How many lives are oppressed, how much joy is erased, by rugged American individualism, and by a culture that says we have to be "nice"?

The answer to all three questions is "All of them." We are trapped in a system that none of us built, but that all of us have helped maintain. White men like me have done the most maintaining, but anytime we allow convenience to overcome curiosity and empathy, we can oppress others.

Cheap pants ordered online may be easy and affordable, but they also drive climate change that causes disasters in the same regions where people are paid exploitive wages to sew those pants together.

We've got to stop pretending that we're a bunch of individual trees when, in fact, we are a forest, a collective whose fortunes are inseparable.

THE SINGLE GREATEST influence in my life toward right-brain embodiment is my friend Hillary McBride. Being her friend has changed me and, in doing so, changed the way I write. The earliest versions of this book talked about emotions, but primarily from the perspective of how our feelings often invisibly shape our cognition. My first drafts offered top-down CBT treatment as the primary way of managing feelings and driving behavioral change.

The only discussion of embodiment was a nod to the role our bodies play in our cognition and our ability to move through the world. But working alongside Hillary didn't just change how I communicate about emotions to others—it changed how I relate to my own feelings. I've joked a few times that I should rename this book "Things I Learned from Hillary."

Hillary has taught me little things, like how to roll my shoulders (seriously). But she also taught me how to listen when my body was trying to tell me I had feelings. She helped me to see the ways in which I've allowed the Man Box to censor what I allow myself to feel, and what I repress in order to do so.

Hillary helped me find a therapist when what all I repressed started to come up, creating panic attacks and creeping anxiety. She taught me how to feel safe telling people not only what I love about them, but what my needs are.

I want to be clear here. Hillary never tried to be my mentor and was never my therapist. Hillary taught me all this with her own actions in friendship. I've watched as this wonderful, weird soul became a part of my friend group, and her example invited us toward more health, self-acceptance, and mutual grace.

This work of listening to my body is helping me to climb out of the Man Box. It's also helping me to discover the alternative to the oppressive niceness I've always lived in. Niceness means avoiding discomfort in ourselves, and avoiding provoking it in others. This fear of discomfort dissuades us from sharing or listening to serious matters of injustice in our world. You can be nice by rolling your shopping cart back to the store while walking past a homeless person without acknowledging that they exist.

Kindness, on the other hand, means taking ownership of your agency in a way that fosters intimacy. Niceness says a black woman can't be angry when she discusses police brutality, but kindness says that supportive listening and acknowledgment of pain is the only way to live in relationship. Niceness is disembodied, while kindness begins in our guts and in our bones. Niceness is a veneer, but kindness goes to the heart.

Embodiment lets us begin to see our own hang-ups and hidden hurts. White people can't be kind to people of color until they are in touch with their bodies enough to not only empathize with suffering, but also take ownership of their own feelings, beliefs, and behaviors. When I, as a white person, am in my body, the fact that people with my skin color are prosecuted less often and receive more lenient sentences than anyone else in this country isn't just uncomfortable, it's a heart-rending injustice. I can be aware of my own shame, and deal with it, without projecting that back onto the person of color who had the courage to describe the truth of how they are treated in our society.

Moving into my body has helped me to have deep friendships with people of all genders, orientations, ethnicities, races, sexual orientations, and disabilities. Because when I am in my body, in touch with the sensations and emotions of living, my friends don't have to take responsibility for my feelings—and I am liberated from the Man Box in a way that tells me I don't have to be their leader or their savior.

I can be for them what Hillary has been for me: a friend. Another tree in an interdependent forest, finding mutual shelter and aid as we create a healthier environment for all.

Tilling Better Soil

*Learning to Take Care of
Our Most Limited Resource*

As my Lyft driver dropped me off at LAX, I read an email that said my friend Rachel Held Evans was dead. In a state of panic and confusion, I called my wife and we sobbed over the phone. I didn't know what to do—about the flight I was about to board, sure, but also about life. What would we do without Rachel?

I am a Christian. But I am the kind whom most American Christians wouldn't call a Christian, and whose community is mostly made up of people who have no faith affiliation. Call us the Church of the Nones. Among our number, Rachel Held Evans is a saint.

People often use the word "saint" in casual conversation to mean someone who is loving and kind. But when I say Rachel is a saint, I mean it in the formal church meaning of the word. Rachel was one of the most gifted authors of her generation, and she used her voice powerfully. She was raised in the heart of the Evangelical church, where the pastors taught that she should love Jesus, minister to the lost and poor, and never, ever teach or hold authority over a man. I can imagine a young Rachel looking at that list and saying, "Well, two out of three ain't bad."

Rachel fought powerfully and tirelessly for the people that Christians often leave behind: women, people of color, LGBTQ folks, and disabled people. She welcomed doubters, skeptics, the grieving, the broken, the torn, and the tired. With seemingly endless compassion and unfailing love for not only the church, but also the Bible and even Christian theology, Rachel grew up among those who said she had no authority, and she went ahead and changed the world anyway.

So when I call her a saint, I am saying that if the Church of the Nones had a path to sainthood, Rachel Held Evans would be canonized immediately. Saint Rachel, patron of broken-hearted doubters. God, how I loved her. And love her still.

Over the years, Rachel (and her husband, Dan) became dear friends to me. She helped get The Liturgists off the ground through her enthusiastic participation. She coached me through the process of writing a book and helped me sign with a speaking agent for the first time. I shouldn't have been surprised. Rachel did that for me, yes, but she did it for dozens of others. Her courage made the space for us to do what we were called to do, to fight for justice and inclusion in a faith that so often doesn't want any form of it.

I was devastated when we lost Rachel. I couldn't function, not professionally or personally. I've lost people I love before, but Rachel, well, that was different. Rachel was the best of us. Rachel was why I still called myself a Christian, and why I still cared about the church. Losing her broke my heart.

I mean that in an obvious way—that I was grieving and hurt. But I also mean it quite literally. My heart broke, physically, under the grief of that loss. The day before Rachel's funeral, I landed in the hospital.

When an obese, middle-aged man walks into the ER and says, "I'm having chest pains," things move fast. They don't hand you a form and ask you to "sit over there." Someone comes to get you immediately, and they take you back and start taking off your clothes. They attach electrodes to your body and before that's all done, a doctor is there asking you questions and checking your vitals. It happens with a speed I've never experienced in a medical setting.

I didn't think I was having a heart attack. I'd woken up with chest pains multiple times overnight. The first crushing wave of pressure that woke me up felt like death, but I'd survived twenty minutes of it and gone back to sleep. Same with the second, and the third, and the fifth. A wise man would have woken his wife in the middle of what seemed like impending death, but I wanted Jenny to be able to rest.

That's right, dear reader. At the moment when I was almost certain I was about to die, I didn't want to bother my wife to tell her I was having a heart attack. Talk about living in the Man Box.

When we got out of bed that morning, I still couldn't catch a breath, and the pressure wouldn't stop. Finally, I got serious. I told Jenny I was having chest pains. Her eyes widened. "Do you think you're having a heart attack?" she said.

"I don't think so," I said, trying to comfort her. "I woke up several times with crushing chest pains last night, but I'm not dead, so I don't think it's a heart attack. I've been on Google, and I think it could be angina, which is a common precursor to a heart attack."

"Oh my God," Jenny said. "I'm calling an ambulance."

When we arrived at the hospital, they took me straight back and started connecting sticky probes all over my hairy body. SPOILER ALERT: When I got home, taking those damn things off looked like the waxing scene from *The 40-Year-Old Virgin*.

I was not having a heart attack, they learned, but my blood test showed some trouble, and my symptoms suggested a heart attack could be imminent. So, my ER visit became a hospital stay with dozens of tests. They really take chest pains in middle-aged men seriously.

With all of this chaos, I ended up missing my flight to Rachel's funeral and watched the whole thing at home on bed rest via a live stream. I was so sad to miss it, but my ticker was plenty provoked by the grief from the watching. Every familiar face of a friend on-screen unleashed another round of chest pressure and stole the breath from my chest.

I was diagnosed with pericarditis, among other things. Pericarditis is an inflammation of the membranes around

the heart. It's most common following respiratory infec-
tions and chest and heart surgeries, and in people in their
seventies and up. They don't often find it in forty-one-
year-old men who've never had surgery and aren't otherwise sick. My cardiologist was incredibly thorough, asking
all about my diet and lifestyle, and damn near getting me
to confess the name of my imaginary girlfriend from third
grade. She crunched the numbers and came up with a theory: chronic stress and too little rest.

MY LIFE HAS changed so much during the process of
writing this book. I was diagnosed with autism. I started
trauma therapy to deal with panic attacks. My daughter
Madison was diagnosed with anorexia (which I share with
her permission, but it is her story to tell). For the first time,
I told my wife and the public that I am a survivor of sexual
assault. And that just scratched the surface of the last year
and a half. It's been a lot to take.

I started writing a book about how I used science to
help me deal with things that happened to me in the past,
and instead had to try to apply those insights to a roller-
coaster ride of joy and grief. And, through all that, I
knew—I mean I absolutely knew—that I was kicking one
important can down the road: my health. I was doing
good work in public advocacy and education. I was working even harder for my mental health. But I asked my poor
body to carry that load by working all the time, and eating
junk so I could feel happy while my life came unglued.

I told myself that I would focus on my diet and exercise

habits as soon as the book was done. But before I could get there, my overworked and undernourished heart finally said enough.

My friends: We are always in process. This book is not about being some finished product, some enlightened soul who transcends pain and loss, and always sends a thoughtful thank-you note. No, this book says *you are a miracle*, because like me you survive all this and worse, and you are still here. This book says you are stronger than you'll ever imagine, even if you feel weak.

As I WAITED to take a cardiac stress test, one of the cardiologists told me that I wasn't going to make it to sixty unless I changed my ways. Another said that physical health was the foundation of mental health, and I had to stop thinking of my brain and body as separate entities.

My primary cardiologist rightly identified me as a workaholic and an emotional eater. She told me that I needed some mentally engaging forms of rest and stress management. She recommended two hobbies: bird-watching and tending a vegetable garden.

So far, the memory of lying in bed with crushing pressure in my chest, imagining it was the final moments of my life, has been a pretty great motivator for change. I eat mostly leafy green vegetables and legumes now, and the practices of watching birds and tending a garden have helped with the stress. The daily rhythms of filling feeders, tweaking their placements, and scrubbing the bird bath fill me with such contentment. I love watching the birds' interactions and noticing how different species have personalities. House finches jockey for position with an aggressiveness that far outweighs their size, while mourning doves have an easy disposition that allows them to grow comfortable with my presence in the garden. Bird-watching has been an unqualified success.

The vegetable garden, less so. Though my tomatoes are thriving, my herbs and carrots all died within a week. My peppers, well, I think they'll make it, but it was quite a rough go. Tending that garden, my hands in soil and water, has taught me so much about what I need to know right now, in this time when my love for my work and the reemergence of old wounds conspired to push my health beyond the breaking point.

I'm learning that thoughtful people are often full of self-doubt. There's even a name for this: imposter syndrome. I know so many artists, poets, scientists, and other true experts who wonder if they are qualified to do their work. Imposter syndrome shows up because research tells us that the same set of skills to do something also are needed to know if you are any good at it. I most often experience imposter syndrome when writing, but the very presence of imposter syndrome should encourage me that I have the potential to write well.

The flip side of imposter syndrome is the Dunning-Kruger effect. Named for the psychologists who developed it, the Dunning-Kruger effect tells us that people with little knowledge on a subject often feel like experts—they lack the knowledge to even know what they don't know. I've often seen teachers reference imposter syndrome to encourage people with self-doubt, while the Dunning-Kruger effect is used to dismiss people as thoughtless rubes. But my vegetable garden is teaching me that we're all gifted people experiencing imposter syndrome sometimes, while blind to our own Dunning-Kruger stylings at others.

So I'm learning to accept that for now I'm good at tomatoes but hopeless when it comes to carrots. There's power in saying "I'm not good at this yet" instead of "I am a failure at this." That's a big deal: It's the focus of the work of Carol Dweck, a researcher who specializes in motivation. She found that the ability to shift from "I'm not good at this" to "I'm not good at this yet" is one of the primary indicators between people who are satisfied with their personal and professional growth and those who are not.

If I give up because I feel like an imposter, I'll never have a garden at all. But if I assume my tomatoes indicate I am brilliant and beyond critique, then I will never have home-grown carrots on my plate.

I've learned something else from the dozens of ficus trees that line my garden in an orderly row. They're all supported by stakes in the ground. Some of them are getting too big and need to have the stakes removed, but others still can't stand on their own—they'll collapse in the first strong wind. Sometimes living things need support to grow, but sometimes that support can end up confining them.

My stake in the ground was the Evangelical Church. For years, it supported me well, but like a cast left on too long, eventually that system started to confine me more than it supported me. When I finally left, I didn't have the strength to stand alone. That new era of freedom was both liberating and terrifying, and I've spent the last few years both growing and coming unglued. That's part of the process.

I'm learning to be patient with myself while old wounds mend. Taking the time to move slowly through life, to learn and reflect, and to give space to my feelings is new work. I will have to limp before I can take a confident stride—and I have to lean on others as I learn to walk on my own. Just like my ficus trees.

I AM LEARNING to tend better soil. In gardening, it's easy to add too much fertilizer or water. You also learn that different plants need different amounts of sunlight, even when they're described as "full sun." Tending a garden means learning to be aware of so many factors, and, well, sometimes it's just beyond me. So it is with tending to our hearts. Different seasons are "hotter" and "drier"; in those times, it's hard to survive, much less grow. But there are also times when the weather is just right, and everything grows beautifully.

There is a season for every kind of plant, and there is a season for every emotion our hearts can grow. Right now, I am in a season with a harvest of sadness, but I know that it will not always be this way. Besides, sadness isn't such a bad crop after all.

So, I am slowing down. The frenetic pace with which I've made work since becoming Science Mike is at an end.

I'm exploring ways to share the sunlight with others, who can use my platform to share what they've learned in their own gardens. I'm changing old habits and getting a little more difficult to reach. I'm pulling up the weeds of social media obsession and revenue forecasting so the things I care about can grow in my heart instead.

I'm creating shelter for myself by carefully curating the media I consume, reading the news less frequently (once a day is just fine, thank you). When I do so, I focus on stories of beauty, joy, and the exploration of truth as much as I do stories of violence, American exceptionalism, and consumerism. Without denying or ignoring the pain, hurt, and loss in the world, I am taking more time to remember why we want to live in the first place.

Most important, I am learning to be careful with the harvest. My first basil plant was so delicious, I clipped its leaves every day to garnish our family's meals. It was great—until the plant died. It didn't have time to restore what it lost, much less grow, with the speed I was clipping.

My god, my heart aches just writing that. Because that basil plant is me. And I bet it's you too. We live in a capitalistic culture that values productivity and output. We treat our own creativity and emotions in the same careless manner that our species treats our forests. We cut what we need, as fast as we need it, paying no attention to how the forests shrink and the climate changes. Soon, we are shocked when we have a nervous breakdown, panic attacks, or anxiety.

We, people of great material resources, clear-cut our hearts to look impressive at work, or advance causes we care about. The misery we so often feel is our bodies and minds trying to tell us something: that we've had enough. It's time to slow down, to rest, to restock, and, yes, to grow some new leaves before we cut any more.

My basil plant is massive now. I've stopped stripping it of its leaves, and wow . . . that plant is practically a shrub.

SADLY, THESE RECENT efforts to improve my health are not my first. When my kids were toddlers, I got so busy trying to be a parent of small children that I let bad eating habits get my weight up to three hundred pounds. While science supports the notion that you can be healthy at any size, my three-hundred-pound self was not healthy, not in any way. So I applied my passion for data-driven behavioral change to get in better shape.

I decided that training for a marathon was the way to go—and I was right. Running forty miles or more per week really did take the weight off, while letting me continue to stress eat. But I was surprised to learn that I really love running. Once I got past the pain in my lungs and the pin-pricks on my skin, my body adapted to the challenge, and my misery turned to elation.

Most of my first book was written in my mind while my feet carried me along the pavement and trails that I liked to run on in the early morning. My favorite trail was a narrow path that crossed some train tracks and ran along a lake before leading deep into the woods. I liked looking up at the tree canopy and watching the sunlight come through the leaves in glorious beams. I loved watching the squirrels. They'd run, as I did, along the ground, but if I startled them, they always seemed to have a "favorite" tree to run up for safety. Often they would run past a closer tree to get to another—which seems like a risky move if I were an actual predator, and not an obese man trying to get healthy.

All those trees come from the same source, the earth, and then stretch toward the sky as individual tree trunks. But, once their branches reach high enough, every tree merges again into a single canopy. Squirrels may have a favorite trunk, but they all climb to the same canopy. So it is with us.

You are a miracle because you have a home tree that you learned to climb well. For me, that tree was Christianity. I have a foothold in the Hebrew Bible, a handhold in what Christians call the New Testament, and, yes, a grip on the words left by my friend Rachel Held Evans, leading me up and into the light.

But as I have grown, the Evangelical tree isn't my home tree anymore. I split time between interacting with mainline Protestants and humanists. The writings of Carl Sagan and Ta-Nehisi Coates propel me toward the light just as well as those from the Christian tradition. I'm learning to explore more of the forest without fear and to count every tree as worthy, as belonging, and, indeed, as sacred.

One day, maybe I'll reach the point where I include myself.

. . .

THE LAST YEAR has been financially ruinous for my family. Even as I wrote in this very book on the value of interdependence and the danger of self-sufficiency, I was watching medical debts pile so high, and our savings sink so low, that I wondered how we'd make rent. In typical fashion, I hid it as long as I could.

Then we ran out of money. The savings I'd spent my adult life building up evaporated. I started looking at options, from returning to IT to driving for Uber. It got to the point where our family couldn't come up with the next rent check that was due.

That's when the bill for my hospital stay came. I laughed. Forty thousand dollars may as well have been $400,000 or $4 million. It was an impossible amount of money. I told a few friends that we were in dire straits, and told the team at The Liturgists that I was going to have to take my work part-time.

Hillary told me she was going to start a GoFundMe, and my body flushed with panic. I delayed and stalled, reminding Hillary that people already send me money every month through the podcasts' listener support programs. But my friends Michael and Caroline jumped in and told Hillary that I'd delay forever. The only way forward, they said, was to just launch it.

So they did, on my forty-first birthday. Michael and I were on tour, and every so often Michael would tell me what the dollar total was at that moment, because I couldn't bring myself to look. The campaign hit a $50,000 goal in less than two days. People sent checks and Venmo payments, and, well, the campaign is only part of what people sent my family. Our debt was erased by the kindness of friends and strangers.

I felt grateful, of course, but also ashamed. Taking care of my family was the core of my identity, and somewhere along the line, my best intentions had gone off track. I felt like a failure. I wondered if I should have stayed in Tallahassee, working in IT, instead of moving to Los Angeles to work as an author and podcaster.

The truth is, what happened to me could happen to anyone. I am careful with money and live a modest life. But that insight didn't offer me any relief, precisely because what happened to me *could* happen to anyone—but not everyone is a popular podcaster with an audience that is happy to help out in a time of need.

I've been stuck in a cycle of asking, "Why me?" I feel like George Bailey in *It's a Wonderful Life*: a basically kind person who was lucky to do work that garnered affection from a community. I am well loved, and well cared for by people all over the world.

But I am also a work in progress. Even as I write the final words of this book, I am struggling to learn and process everything in its pages. I am a mess, but one that is full of love, propelled to generosity, and learning to embrace the full range of my human experience. Sometimes I get so lost in surviving that I forget to actually live. I know this because I've gone from thinking I could write a book about growth and transformation, to sharing a real-time sampling of loss and suffering in the time it took to write these words.

But in doing so, I have made deeper and more genuine friendships than I've ever had. I have learned to love the feelings I once feared. And, finally, I am learning that my penchant for self-sufficiency is a dangerous and self-indulgent habit.

I am learning every day to stop living a performative life. I'm starting to accept that I am who I am, while trusting my potential to grow and change as needed—my heritage as part of the amazing web of life on our planet.

I am a miracle, and a pain in the ass.

Friend, so are you. Your story isn't mine. You've followed a different path, with its own highs and lows. You've faced challenges, and you've made it through. I know that because you're here, reading these words right now.

I hope you learn to see the miracle, to claim your rightful place among wonders like supernovas and photosynthesis. But I also hope you understand that the things that drive you crazy about yourself are there to protect you, to help you, and to drive you toward the change you'll need to make it through the next rough patch in your life.

Most of all, I hope you learn to accept the miracle that is every person you've ever met, and ever will meet. When they're a pain in your ass—well, now you know that they're just trying to survive too.

Acknowledgments

This book would be a very different book if my life hadn't fallen apart while I wrote it. But the way in which I recovered from the pain of these last two years has been shaped by my friend and cohost Hillary McBride in a dramatic way. Thank you for showing me how to love my feelings, and thank you for never giving up on me.

In that way, there are a number of other people who helped me survive—and in doing so, made this book possible. So . . .

Jenny, my lover, my lady, thank you for always protecting me, caring for me, and keeping my shoes on my feet.

Madison, thank you for showing me how to struggle while still having grace and compassion.

Macey, thank you for bringing light when days are dim.

Ruth McHargue, thank you for making sure I have always known I am loved.

Mike McHargue, thank you for showing me how to walk with a wound.

Melissa McHargue Keahey, thank you for showing me how to hold on to what really matters.

Vishnu Dass, we have bonds forged on the astral plane. Thank you for making me laugh on the darkest days.

William Matthews, thank you for showing me how to show up.

Lisa Gungor: You are more than a friend to me. You are a sister, a companion, and a fellow aching heart who yearns for the world to be less brutal.

Christy and Link, thank you for having faith in me when I had none.

Jessie and Rhett, thank you for giving me courage when I was afraid.

Emily and Stefano, thank you for teaching me about eating plants. It saved my life.

Caroline and Jayden, thank you for showing me how to find home on an adventure.

Pete Holmes, do you have to be so FUNNY, so BRILLIANT, and so KIND? Save some stat points for other people!

Joyce Derrickson, I don't know where you found a life raft with wine and cheese in it, but thanks for dragging me out of the surf and immediately handing me a glass.

Austin Channing Brown, you are an even more powerful friend than you are a powerful writer.

Kevin Garcia, thank you for being my safe space.

Micky ScottBey Jones, thank you for showing me how to be a friend instead of a robot.

Corey Pigg, thanks for keeping me on my feet.

Jamie Lee Finch, I am overjoyed our child selves found each other.

Tori Williams Douglass, I am thankful to be your white friend.

Shannon Dingle, you will never know how much you did to help me understand and accept my autism.

Steve Fortunato, we ran different ways from the same start, but I'm so glad our paths converged on the other side.

André Henry, I like to do the work with you, yes, but I like to grab a drink with you as well. Thank you for reminding me why we fight, and why I still want to be a Christian.

Audrey Assad, your stories helped me to understand myself and to begin to heal.

Ryan O'Neal, friend, if you ever stop writing music, I can't write any more books. Thank you for inspiring me.

Colleen Pettit, I can't wait to meet the new one!

Stratton Glaze, you are the scaffold I lean on, even when you don't know it.

Gregg Nordin, you are almost as funny as you are kind. I am so proud to know you.

Tanner Hearne, thanks for making the world a better place through careful planning and precise execution.

Victory Palmisano, you see people and make them feel seen. Thanks for showing me how to do that well.

Caitlin Hermstad, working for you is the smartest thing I've ever done.

Bradley Grinnen, let's stay alive together, eh? These hearts have more beating to do.

Sarah Heath, thanks for introducing me to the "messy middle."

Brent Kredel, I would have given up without you.

Christopher Ferebee, my friend, I would have never started this book if not for you. Thanks for reminding me an author writes books.

Tina Constable, I could not imagine a better publishing partner. You are a marvel, and a kind soul.

Derek Reed, for the first time in this entire section I am at a loss for words. You are one of the most gifted people I have ever met, and you've refined that gift through hard work. Thank you for helping me find the right words, even when I get lost in the woods.

Finally, to each and every one of you who went to Go-FundMe or Venmo and made my hospital bills disappear: I understand how George Bailey is the richest man in town because of you. You may think my work has impacted you, but you saved my life and I will never forget it.

Appendix A

A Tour of Your Brain

Back in chapter 2, I described the brain as a burrito. But the brain is so much more complicated than that, or the structures I was able to tell you about in that chapter. So please enjoy this extended tour of your own brain, to learn about the major players in every choice you make (and especially in the ones you don't).

The Brain Stem (or Reptile Brain)

The oldest structures in your brain are all part of the brain stem. These structures are similar to the ones found in our ancestors who were fish or amphibians, and they play similar roles for us today that they played then.

Your *medulla oblongata* coordinates actions in your respiratory and circulatory system—breathing, heartbeat, and blood pressure. It also makes you vomit if it thinks there's something toxic in your body's systems.

Your *pons* handles your posture, facial expressions, chewing, bladder control, and tears.

Your *cerebellum* looks like a smaller brain attached to the bottom of your brain. It handles all your voluntary movements, and also helps you keep your balance.

All these structures together are called the *hindbrain*, and they're tasked with the most basic, essential survival functions. Above them is the *midbrain,* a small patch of neural tissue that plays a role in almost everything your brain does.

The Limbic System (or Mammal Brain)

Your limbic system sits above your brain stem—a brain stacked on top of another brain. I like to think of it as the "dog brain," because it's responsible for almost all behaviors that humans have in common with dogs: eating, having sex, fighting, fleeing from threats, etc. So while your brain stem may "keep the lights on" by managing many of your body's essential systems, the limbic brain is responsible for the drives and impulses that keep you alive.

Because of that ancient survival orientation, a lot of the behaviors and feelings you struggle with in life originate in your limbic system. Your limbic system is the part of your brain that craves calories—even at ten P.M. It's the part of you that wants social validation, and produces anxiety when you don't get it. If you find yourself in some kind of internal conflict, your limbic system is very likely to be involved—and likely to be the part of you that wants to do something other parts of your brain deem irresponsible or short-sighted.

The following are some of the most prominent structures in the limbic system:

Your *thalamus* is located near the center of your brain. I like to think of it as the brain's Grand Central Terminal, where information from your body's senses most often gathers before being routed off to other parts of the brain for processing.

Your *hippocampus* plays an essential role in working memory—the short-term memory we use for daily tasks. Its name is Greek for "seahorse," which is fitting because it looks so much like one. Damage to the hippocampus often causes amnesia.

Your *corpus collosum* is a thick communication channel that connects the two hemispheres of the brain.

Finally, your *amygdala* coordinates fear and anger in your brain. These are the two most powerful human emotions, able to override other thoughts and feelings happening elsewhere in the brain. When a movie scares you or a careless driver angers you, your amygdala is sending a signal to every other part of your brain: We are in danger.

Like many brain structures, your hippocampus and amygdala are mirrored structures in the left and right hemispheres of your brain.

I have to mention as well the *cingulate cortex*, a region of the brain between the limbic system and the neocortex. Of particular note is the anterior cingulate cortex, which coordinates our feelings of compassion, empathy, and intimacy. It's the cuddliest part of your brain.

The Neocortex (or Human Brain)

Finally, we arrive at the wrinkled-up tortilla wrapped around your limbic system and parts of the hindbrain: the neocortex. The neocortex is home to all the parts of your brain that are most uniquely human, which is why I'm calling it the "human brain" in our three-part brain model here, even though the brainstem, the limbic system, and the neocortex are obviously all part of the human brain.

The neocortex is divided into four lobes: the frontal lobe, the temporal lobe, the parietal lobe, and the occipital lobe. The frontal lobe is at the front of the brain, the occipital at the rear, the temporal lobe(s) on the left and right sides, and the parietal roughly at the top.

When I've described what the different parts of the brain do so far, I've simplified their functions for the sake of accessibility. But I want to note that when we talk about the neocortex, you could fill a book with the functions of each lobe, and you'd also find their duties overlap significantly. Keep that in mind as I talk about what each lobe does. In something as structurally complex as the neocortex, any explanation we laypeople can understand sacrifices clinical fidelity.

First up is the frontal lobe. Anytime you choose to do or think something, the frontal lobe is involved. That's often called executive function. There are some notable structures in the frontal lobe that we'll talk about often over the course of this book.

The *prefrontal cortex* is often called the brain's CEO.

It's a quarter-size patch of tissue at the front of each of the brain's two hemispheres. The prefrontal cortex is the seat of rational and analytical thought. It doesn't know everything that's happening in the brain, but often acts as a mediator or decider when brain networks are in conflict.

We talked about the prefrontal cortex often in this book, but I want to make something clear. The CEO description of the prefrontal cortex is helpful for understanding executive function, but it overstates the prefrontal cortex's power. If the parts of the brain are animals and employees in a zoo, then perhaps the prefrontal cortex is better thought of as a zookeeper. CEOs can order employees to do just about anything, but a zookeeper has to respect the power of the lions, wolves, and elephants in her zoo. A similar relationship exists between the more rationally oriented focus of the prefrontal cortex and the more emotional focus of the limbic brain. When you try to calm yourself during a fear response, it's less your rational brain ordering your emotional brain to calm down, and more like a zookeeper soothing an upset animal. Notably, the relationship between the two over the years will impact how well this process goes.

The prefrontal cortex itself is divided into distinct regions as well. One notable one for our purposes is the *orbitofrontal cortex*—named for the orbit of the eye. Your orbitofrontal cortex plays a key role in your capacity to weigh the consequences of potential actions. This part of your brain is obsessed with predicting the future.

The orbitofrontal cortex is deeply wired to regions in

your amygdala, and is believed to allow the more primitive emotions like fear to become something more nuanced, like anxiety.

Interestingly, the prefrontal cortex doesn't develop as quickly as other parts of the brain. The prefrontal cortex (including the orbitofrontal cortex) is underdeveloped in teenaged brains, leading to a deficit in executive function when compared with adults. Meanwhile, a region called the *nucleus accumbens* matures quickly and allows you to seek pleasure and reward. Teenagers have a neurologically reduced capacity to weigh the consequences of their actions, but powerful emotions and pleasure rewards. I don't know about you, but that makes me feel a lot better about my memories of high school.

At the back of the prefrontal cortex, bordering the parietal lobe, is the *primary motor cortex*. This part of the brain handles voluntary movement. When you turn the pages of this book, that's the primary motor cortex doing its thing.

Next up is the parietal lobe. It is responsible for creating your sensations of touch and spatial awareness. When you feel the wind on your cheek, or the gentle touch of a lover, your parietal lobe is transforming nerve signals into sensation. At the very front of the parietal lobe is a strip of tissue called the *primary somatosensory cortex*, and it's responsible for the sense of touch coming from all the parts of your body.

This part of the brain is well mapped. Because of that, we understand that the brain devotes more processing power to your hands and face than to the rest of your body combined (though, thanks to the brain's plasticity, that can change for people who don't have hands). Our mouth and fingers are particularly heavily weighted—which makes babies' finger and thumb sucking make sense. There're few ways an infant can create so much neural stimulation as a feedback loop between her fingers and her mouth.

Below the parietal lobe and behind the frontal lobe is the temporal lobe. All the brain's lobes are mirrored across the brain's hemispheres, but based on the temporal lobe's placement, its two halves aren't in contact with each other at the brain's surface. The temporal lobe is heavily involved in memory formation. It's also involved in your ability to process sounds, particularly speech, pitch, and rhythm. The temporal lobe is your brain's musician.

At the back of your brain is the occipital lobe. It's almost entirely devoted to processing visual stimulus (though it gets re-tasked whenever someone can't see with their eyes). When compared with most animals, humans have an incredible number of neurons devoted to visual processing, which is why we have one of the most sophisticated visual systems on earth today.

When I laid out many of the brain's structures for you in chapter 2, I noted that there's considerable ambiguity when we talk about what any one region of the brain does. That's because your brain's structures organize into networks. In the "zoo" analogy, I think of these networks as relationships between the different animals and employees in a zoo. Just like you can have more than one friendship, or even more than one circle of friends, the brain's structures can be involved in more than one network.

In the same way, just like your personality may change when you are around different friends, your brain's structures have different "personalities" depending on which network they're interacting with at a given moment.

For example, when you are awake and alert, many of your brain's structures, particularly in the thalamus, frontal lobe, and parietal lobe, form something called the *default mode network*. Sometimes referred to as "task negative," the default mode network is active when you are thinking but not necessarily doing. For some time, neuroscientists believed the default mode network was basically a daydreaming network, but more recently they have discovered it's active very often, and can play a role in task execution, but especially in goal setting and task planning.

Appendix B

Generations Are Bullshit (but We Use Them Because They're Easy to Say)

In media and casual conversation, generational labels (like "millennial" or "baby boomer") often get confused with sociology. This annoys me, because those terms aren't based in actual research. Here's where we got those terms, and why we need to do a better job framing them.

When we talk about generations in the United States, we usually talk about four groups: the Silent Generation, the baby boomers, Generation X, millennials (also called Generation Y), and the iGeneration (also called Generation Z or the post-millennials). Each generation covers roughly twenty years of births.

The Silent Generation fought in World War II and built the postwar United States. The baby boomers were born in the rapid economic expansion that happened in the United States following World War II. They danced at Woodstock and chased money in the Wall Street boom of the 1980s.

Generation X, my generation, happened during a decline in the birth rate. We listened to grunge music as our parents divorced, forever stuck in the shadow of the much larger generation that preceded us. We're famously cynical and apathetic.

The millennials represent another birthrate boom, and thanks to their numbers, they quickly overtook Generation X in media attention. If you pay attention, most issues in American media are defined as boomers vs. millennials, with nary a mention of Gen X. Millennials are "digital natives," meaning they grew up surrounded by devices with Internet connections.

Finally, we have the iGeneration. They've grown up not just with the Internet, but with the Internet in their pocket. They're also the only American generation still being born.

Most of this is common knowledge, but I wanted to offer a refresher because we talked about generations a lot in chapter 5. And as a lover of the sciences, I also want to offer some important nuance.

Generations sound so official and scientific that most people assume that these categories come from social science. They don't. If you read the research published by social scientists, you won't see any of these generation names. Social scientists often do a type of research called cohort studies. Cohort studies involve studying a group of people with a common characteristic or experience. It's one of the fundamental tools that scientists use to understand people.

Cohort studies are complex and nuanced, to the point they're hard for nonacademics to understand. So journalists will read them, find common threads, and then write about them. If you're a journalist, tying a lot of cohort research together and coining it with a catchy name is a great way to move your career forward. That's how we get to magazine covers that say "Meet the Millennials."

These terms don't come from scientists but from journalists trying to help us understand what scientists are finding out about people. I think that's good! The problem comes when we assume that these labels have fixed definitions. But different journalists (and the scientists using these terms in interviews once most people know them) may mean something different when they say "baby boomer." They may not be talking about a full twenty years, or if they are, they may use a different starting and ending date than someone else.

No one is right or wrong when that happens—these terms aren't officially defined. They're just a shorthand used to summarize (and often overgeneralize) the findings of cohort studies.

Notes and Further Reading

This book followed a wandering path from conception to publication. Some works played broad, inspirational roles in the content of this book, while others are specific citations. Both are listed below, in the context of the chapter they most influenced.

Chapter 1

These books were influential in my exploration of the "Divided Self" and are well worth reading to explore the idea more deeply: *The Honest Truth About Dishonesty* by Dan Ariely; *The Power of Habit* by Charles Duhigg; and *The Happiness Hypothesis* by Jonathan Haidt.

8 **exactly what the BBC found** "The Science of the Young Ones: Priming," BBC, https://www.youtube.com/watch?v=5g4_v4JStOU.

9 **we associate with being polite** Dijksterhuis and van Knippenberg, "Seeing One Thing and Doing Another: Contrast Effects in Automatic Behavior," *Journal of Personality and Social Psychology* 75, no. 4 (October 1998): 862–71.

9 **Another study indicated that people** Wheeler and deMarree,

"Multiple Mechanisms of Prime-to-Behavior Effects," *Social and Personality Psychology Compass* 3, no. 4 (July 2009): 566–81.

9 **research is showing us** Ryota Kanai, Tom Feilden, Colin Firth, and Geraint Rees, "Political Orientations Are Correlated with Brain Structure in Young Adults," *Current Biology* 21, no. 8 (April 2011): 677–80.

Chapter 2

If you want to dig deeper into understanding the human brain, I recommend these books: *The Brain* by David Eagleman and *Phantoms in the Brain* by V. S. Ramachandran.

For a look at other ways evolution has produced intelligence (which may help you better understand your own), try *Other Minds* by Peter Godfrey Smith and *Are We Smart Enough to Know How Smart Animals Are?* by Frans De Waal.

For an incredibly detailed look at anger (specifically, the out-of-control manner we call rage), try *Why We Snap* by R. Douglas Fields.

Chapter 3

For a look at compulsions in the context of hoarding, check out *Stuff* by Randy O. Frost and Gail Steketee.

The mechanisms in our brains that create compulsions can be hijacked to serve us. To learn more, read *Nudge* by Richard H. Thaler and *Stick with It* by Sean Young, Ph.D.

46 **a zoologist named Niko Tinbergen** "Nikolaas Tinbergen, Biographical," NobelPrize.org, October 14, 2019, https://www.nobelprize.org/prizes/medicine/1973/tinbergen/biographical.

52 **Ms. Kardashian uses makeup** Theresa E. DiDonato Ph.D., "5 Research-Backed Reasons We Wear Makeup," *Psychology Today*, February 6, 2015, https://www.psychologytoday

.com/us/blog/meet-catch-and-keep/201502/5-research-backed
-reasons-we-wear-makeup.

56 **70 percent, and possibly higher** Marie-Ève Daspe et al.,
"When Pornography Use Feels Out of Control: The Mod-
eration Effect of Relationship and Sexual Satisfaction,"
Journal of Sex & Marital Therapy 44, no. 4 (2018), DOI:
10.1080/0092623X.2017.1405301.

56 **29 percent of their users are women** "2018 Year in Review,"
Pornhub Insights, https://www.pornhub.com/insights/2018
-year-in-review.

56 **Younger women are much more likely to look at porn**
Cassie Murdoch, *Mashable*, April 12, 2017, https://mashable
.com/2017/04/12/youporn-women-porn-data/#rot9TEn
_.mqr.

57 **researchers compared the brains of control** Valerie Voon
et al., "Neural Correlates of Sexual Cue Reactivity in Individuals with and Without Compulsive Sexual Behaviours,"
PLOS One 9, no. 7 (July 11, 2014).

57 **that this compulsion has serious consequences** Dennis
Thompson, "Study Sees Link Between Porn and Sexual
Dysfunction," WebMD, May 12, 2017.

58 **groups also search for porn more** Andrew L. Whitehead
and Samuel L. Perry, "Unbuckling the Bible Belt: A State-
Level Analysis of Religious Factors and Google Searches for
Porn," *The Journal of Sex Research* 55, no. 3 (March–April
2018).

58 **abstinence-only sex education put forward** Kathrin Stanger-
Hall and David W. Hall, "Abstinence-Only Education and
Teen Pregnancy Rates: Why We Need Comprehensive Sex
Education in the U.S.," *PLOS One* 6, no. 10 (October 14,
2011).

58 **women are unable to achieve orgasms** "7 Factors Affecting
Orgasms in Women," *Psychology Today*, April 28, 2014.

61 **psychologist named Bruce K. Alexander** B. K. Alexander,
"Drug Use, Dependence, and Addiction at a British Colum-

bia University: Good News and Bad News," *Canadian Journal of Higher Education* 15 (1985): 77–91.

Chapter 4

Readers can find a practical application of AEDP in *Living Like You Mean It* by Ron Frederick.

The relationship between trauma and our bodies is well explored in *The Body Keeps the Score* by Bessel van der Kolk.

84 **don't experience time slowing** Chess Stetson, Matthew O. Fiesta, and David M. Eagleman, "Does Time Really Slow Down During a Frightening Event?" *PLOS One* 2, no. 12 (December 12, 2007).

85 **530 million neurons in a dog's brain** Débora Jardim-Messeder et al., "Dogs Have the Most Neurons, Though Not the Largest Brain: Trade-Off Between Body Mass and Number of Neurons in the Cerebral Cortex of Large Carnivoran Species," *Frontiers in Neuroanatomy* (December 12, 2017).

Chapter 5

The different modes of our neurocognitive systems are explored in *Thinking Fast and Slow* by Daniel Kahneman. *The Sentient Machine* by Amir Husan helped me frame my thoughts on how machine learning may shape our world over time. *Reclaiming Conversation* by Sherry Turkle offers a comprehensive view on mitigating digital media in our relationships. *Why We Sleep* by Matthew Walker told me just how bad chronic sleep deficits are for people. If you want to improve your defenses against misinformation on the social web, try *A Field Guide to Lies* by Daniel J. Levitin.

108 **moral indignation goes "viral"** Takuya Sawaoka and Benoît Monin, "The Paradox of Viral Outrage," *Psychological Science* 29, no. 10 (October 2018): 1665–78. doi:10.1177 /0956797618780658.

116 **Here's a quote from Jean M. Twenge** Jean M. Twenge, "Have Smartphones Destroyed a Generation?" *The Atlantic*, September 2017.

119 **experiment, this time involving bilingual people** Renata F. I. Meuter and Alan Allport, "Bilingual Language Switching in Naming: Asymmetrical Costs of Language Selection," *Journal of Memory and Language* 40, no. 1 (January 1999): 25–40.

119 **stick to the same type of task** Robert Rogers and Stephen Monsell, "The Costs of a Predictable Switch Between Simple Cognitive Tasks," *Journal of Experimental Psychology: General* 124, no. 2 (June 1995): 207–31.

120 **executive function has two distinct stages** Joshua S. Rubinstein, David E. Meyer, and Jeffrey E. Evans, "Executive Control of Cognitive Processes in Task Switching," *Journal of Experimental Psychology: Human Perception and Performance* 27, no. 4 (2001): 763–97.

Chapter 6

How We Talk by N. J. Enfield, *Language at the Speed of Sight* by Mark Seidenberg, and *The Influential Mind* by Tali Sharot helped me explore the relationship between language and our brains.

To learn more about depression, I recommend *Lost Connections* by Johann Hari.

Cognitive Behavioral Therapy, 2nd ed., by Judith Beck is a definitive look into CBT.

127 **when the Columbine High School shooting happened** Gillian Brockell, "Bullies and Black Trench Coats: The Columbine Shooting's Most Dangerous Myths," *Washington Post*, April 20, 2019.

128 **word for the color blue** Fiona MacDonald, "There's Evidence Humans Didn't Actually See Blue Until Modern Times," *Science Alert*, April 7, 2018.

Chapter 7

My search for the most contemporary understanding of consciousness led me through *The Mind Club* by Daniel M. Wegner and Kurt Gray, *The Secret Life of the Mind* by Mariano Sigman, and *The Tides of Mind* by David Gelernter. My belief that an "individual self" is an illusion was shaped in part by *The Knowledge Illusion* by Steven Sloman and Philip Fernbach.

Readers looking to learn more about attachment theory should start with *Attached: The New Science of Adult Attachment and How It Can Help You Find—and Keep—Love* by Amir Levine.

For more on the life and work of John Bowlby, Jeremy Holmes wrote *John Bowlby and Attachment Theory*.

Chapter 8

My perspective on the Enlightenment and the brain was first forged by *The Master and His Emissary* by Iain McGilchrist. *If Our Bodies Could Talk* by James Hamblin helped me form a concrete view of our body's systems.

To better understand the body's and brain's roles in our emotions, I read *A General Theory of Love* by Thomas Lewis, Fari Amini, and Richard Lannon.

165 **global neuronal workspace (GNW) model** S. Dehaene, J. P. Changeux, and L. Naccache, "The Global Neuronal Workspace Model of Conscious Access: From Neuronal Architectures to Clinical Applications," in S. Dehaene and Y. Christen (eds.), *Characterizing Consciousness: From Cognition to the Clinic?*, Research and Perspectives in Neurosciences (Berlin and Heidelberg: Springer, 2011).

166 **model called integrated information theory** Giulio Tononi, "An Information Integration Theory of Consciousness," *BMC Neuroscience* 5 (November 2, 2004).

170 **was labeled the Man Box** Brian Heilman, Gary Barker, and

Alexander Harrison, *The Man Box: A Study on Being a Young Man in the U.S., U.K., and Mexico: Key Findings*, Unilever, 2017.

Chapter 9

Though their influences will likely be hard to connect by the reader, this chapter was born out of a trip to the hospital soon after reading *The Power of Meaning* by Emily Esfahani Smith, *The Upward Spiral* by Alex Korb, *The Nature Fix* by Florence Williams, and *How Enlightenment Changes Your Brain* by Andrew Newberg and Mark Waldman.

More on Carol Dweck's groundbreaking work on a fixed versus a growth mindset can be found in her book *Mindset: The New Psychology of Success*.

Index

About the Author

MIKE MCHARGUE (better known as Science Mike) is an author, podcaster, and speaker who travels the world helping people understand themselves and their world using science. His bestselling debut book, *Finding God in the Waves*, is a firsthand account of using science to navigate faith transitions.

Mike hosts *Ask Science Mike*, a weekly question-and-answer podcast helping hundreds of thousands explore the questions they've always been afraid to ask. He also cohosts *The Liturgists Podcast* with his friends Michael Gungor, Hillary McBride, and William Matthews. With more than a million downloads per month, *The Liturgists Podcast* is reshaping how the spiritually homeless and frustrated relate to God.

Mike lives in Los Angeles, California, with his wife, Jenny, and daughters, Madison and Macey.

To learn more about Mike, or to invite him to speak at your organization or event, visit www.mikemchargue .com.